T0211911

Models of Horizontal Eye Movements

Part I: Early Models of Saccades and Smooth Pursuit

Models of Horizontal Eye Movements, Part I: Early Models of Saccades and Smooth Pursuit

John D. Enderle

ISBN: 978-3-031-00514-5   paperback
ISBN: 978-3-031-01642-4   ebook

DOI 10.1007/978-3-031-01642-4

A Publication in the Springer series
SYNTHESIS LECTURES ON BIOMEDICAL ENGINEERING

Lecture #34
Series Editor: John D. Enderle, *University of Connecticut*
Series ISSN
Synthesis Lectures on Biomedical Engineering
Print 1930-0328   Electronic 1930-0336

Synthesis Lectures on Biomedical Engineering

Editor

John D. Enderle, *University of Connecticut*

Lectures in Biomedical Engineering will be comprised of 75- to 150-page publications on advanced and state-of-the-art topics that spans the field of biomedical engineering, from the atom and molecule to large diagnostic equipment. Each lecture covers, for that topic, the fundamental principles in a unified manner, develops underlying concepts needed for sequential material, and progresses to more advanced topics. Computer software and multimedia, when appropriate and available, is included for simulation, computation, visualization and design. The authors selected to write the lectures are leading experts on the subject who have extensive background in theory, application and design.

The series is designed to meet the demands of the 21st century technology and the rapid advancements in the all-encompassing field of biomedical engineering that includes biochemical, biomaterials, biomechanics, bioinstrumentation, physiological modeling, biosignal processing, bioinformatics, biocomplexity, medical and molecular imaging, rehabilitation engineering, biomimetic nano-electrokinetics, biosensors, biotechnology, clinical engineering, biomedical devices, drug discovery and delivery systems, tissue engineering, proteomics, functional genomics, molecular and cellular engineering.

Models of Horizontal Eye Movements, Part I: Early Models of Saccades and Smooth Pursuit
John D. Enderle
2010

Biomedical Technology Assessment: The 3Q Method
Phillip Weinfurt
2010

Strategic Health Technology Incorporation
Binseng Wang
2009

Phonocardiography Signal Processing
Abbas K. Abbas, Rasha Bassam
2009

Intermediate Probability Theory for Biomedical Engineers
John D. Enderle, David C. Farden, Daniel J. Krause
2006

Basic Probability Theory for Biomedical Engineers
John D. Enderle, David C. Farden, Daniel J. Krause
2006

Sensory Organ Replacement and Repair
Gerald E. Miller
2006

Artificial Organs
Gerald E. Miller
2006

Signal Processing of Random Physiological Signals
Charles S. Lessard
2006

Image and Signal Processing for Networked E-Health Applications
Ilias G. Maglogiannis, Kostas Karpouzis, Manolis Wallace
2006

Models of Horizontal Eye Movements

Part I: Early Models of Saccades and Smooth Pursuit

John D. Enderle
University of Connecticut

SYNTHESIS LECTURES ON BIOMEDICAL ENGINEERING #34

ABSTRACT

There are five different types of eye movements: saccades, smooth pursuit, vestibular ocular eye movements, optokinetic eye movements, and vergence eye movements. The purpose of this book is focused primarily on mathematical models of the horizontal saccadic eye movement system and the smooth pursuit system, rather than on how visual information is processed. A saccade is a fast eye movement used to acquire a target by placing the image of the target on the fovea. Smooth pursuit is a slow eye movement used to track a target as it moves by keeping the target on the fovea. The vestibular ocular movement is used to keep the eyes on a target during brief head movements. The optokinetic eye movement is a combination of saccadic and slow eye movements that keeps a full-field image stable on the retina during sustained head rotation. Each of these movements is a conjugate eye movement, that is, movements of both eyes together driven by a common neural source. A vergence movement is a non-conjugate eye movement allowing the eyes to track targets as they come closer or farther away.

In this book, early models of saccades and smooth pursuit are presented. The smooth pursuit system allows tracking of a slow moving target to maintain its position on the fovea. Models of the smooth pursuit have been developed using systems control theory, all involving a negative feedback control system that includes a time delay, controller and plant in the forward loop, with unity feedback. The oculomotor plant and saccade generator are the basic elements of the saccadic system. The oculomotor plant consists of three muscle pairs and the eyeball. A number of oculomotor plant models are described here beginning with the Westheimer model published in 1954, and up through our 1995 model involving a 4th order oculomotor plant model. The work presented here is not an exhaustive coverage of the field, but focused on the interests of the author. In Part II, a state-of-art model of the saccade system is presented, including a neural network that controls the system.

KEYWORDS

smooth pursuit, saccade, main sequence, time-optimal control, system identification

Contents

Acknowledgments

I wish to express my thanks to William Pruehsner for drawing many of the illustrations in this book, and Kerrie Wenzler and Gresa Ajeti for editorial assistance.

John D. Enderle
February 2010

CHAPTER 1

Introduction

1.1   INTRODUCTION

The visual system is our most important sensory system. It provides a view of the world around us captured with receptors in the eyeball that is transmitted to the central nervous system (CNS). The eye movement or oculomotor system is responsible for movement of the eyes so that images are clearly seen. The oculomotor system also responds to auditory and vestibular sensory stimuli. In this book, a qualitative and quantitative description of the saccadic and smooth pursuit eye movement systems are presented for horizontal movements of the eye. A saccadic or fast eye movement involves quickly moving the eye from one image to another image. This type of eye movement is very common, and it is observed most easily while reading; that is, when the end of a line is reached, the eyes are moved quickly to the beginning of the next line. Saccades are also used to locate or acquire targets. Smooth pursuit is a slow eye movement used to track an object as it moves by keeping the eyes on the target. In addition to these two movements, the eye movement system also includes the vestibular ocular movement, optokinetic eye movement, and vergence movement. These three eye movements will be briefly described in this chapter. Vestibular ocular movements are used to maintain the eyes on the target during head movements. Optokinetic eye movements are reflex movements that occur when moving through a target-filled environment or to maintain the eyes on target during continuous head rotation. The optokinetic eye movement is a combination of saccadic and slow eye movements that keeps a full-field image stable on the retina during sustained head rotation. Each of these four eye movements is a conjugate eye movement, that is, movements of both eyes together driven by a common neural source. Vergence eye movements use nonconjugate eye movements to keep the eyes on the target. If the target moves closer, the eyes converge — farther away, they diverge. Each of these movements is controlled by a different neural system, and all of these controllers share the same final common pathway to the eye muscles.

Each eye can be moved within the orbit in three directions: vertically, horizontally, and torsionally. These movements are due to three pairs of agonist - antagonist muscles. These muscles are called antagonistic pairs because their activity opposes each other and follows the principle of reciprocal innervation. Shown in Fig. 1.1 is a diagram illustrating the muscles of the eye, optic nerve and the eyeball. We refer to the three muscle pairs and the eyeball as the oculomotor plant, and the oculomotor system as the oculomotor plant and the neural system controlling the eye movement system.

At the rear of the eyeball is the retina, as shown in Fig. 1.2. Regardless of the input, the oculomotor system is responsible for movement of the eyes so that images are focused on the central

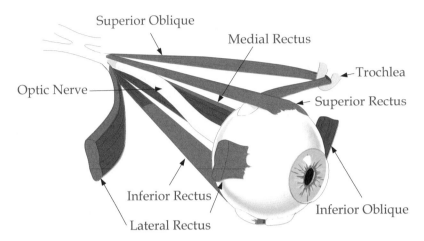

Figure 1.1: Diagram illustrating the muscles, eye ball, and the optic nerve of the right eye. The left eye is similar except the lateral and medial rectus muscles are reversed. The lateral and medial rectus muscles are used to move the eyes in a horizontal motion. The superior rectus, inferior rectus, superior oblique, and inferior oblique are used to move the eyes vertically and torsionally. The contribution from each muscle depends on the position of the eye. When the eyes are looking straight ahead, called primary position, the muscles are stimulated and under tension.

one-half degree region of the retina, known as the fovea. Lining the retina are photoreceptive cells that translate images into neural impulses. These impulses are then transmitted along the optic nerve to the central nervous system via parallel pathways to the superior colliculus and the cerebral cortex. The fovea is more densely packed with photoreceptive cells than the retinal periphery; thus a higher resolution image (or higher visual acuity) is generated in the fovea than the retinal periphery. The purpose of the fovea is to allow us to *clearly* see an object, and the purpose of the retinal periphery is to allow us to *detect* a new object of interest. Once a new object of interest is detected in the periphery, the system redirects the eyes to the new object.

A qualitative description of the each eye movement is provided in the next section. In Chapter 2, models of the smooth pursuit system are presented. After this, chapters on models of the saccade system are presented, including the saccade generator on the basis of anatomical pathways and control theory. The literature on the saccade system is vast, and thus this presentation is not exhaustive but rather a representative sample from the field and the interest of the authors.

1.2   SACCADES

One of the most successfully studied systems in the human is the oculomotor or eye movement system. Some of the reasons for this success are the relative ease in obtaining data, the simplicity

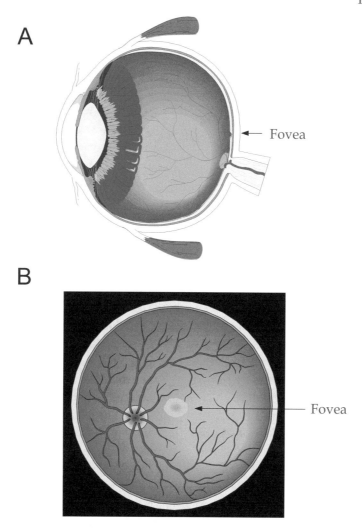

Figure 1.2: (A) Diagram illustrating a side-view of the eye. The rear surface of the eye is called the retina. The retina is part of the central nervous system and consists of two photoreceptors, rods and cones. (B) Front view looking at the rear inside surface (retina) of the eye. The fovea is located centrally and is approximately 1 mm in diameter. The oculomotor system maintains targets centered on the fovea.

of the system, and the lack of feedback during dynamic changes in the system. A saccade is a fast eye movement that involves quickly moving the eyes from one target or image to another. The word saccade originated from the French word *saquer*, which means to jerk the reins of a horse. A saccade is a very quick and jerky movement of the eye from one target to another.

The eye muscles are among the fastest in the human body, with a 10° saccade taking only 50 ms. The saccadic system can be thought of as a targeting system that is concerned only with accurate and swift eye movements from one target to another without concern for the information swept across the retina during the eye movement. During a saccade, the visual system is turned off. After the saccade is complete, the system operates in a closed-loop mode to ensure that the eyes reached the correct destination. Information from the retina and muscle proprioceptors is used to correct any error between the desired and current eye position. The saccade system operates in a closed-loop mode to reduce this error to zero with a corrective saccade. One possible explanation of the operation of the neural control of saccades is that the saccadic neural controller is an open-loop time-optimal system using an internal closed-loop controller (Zhou et al., 2009; Enderle, J., 2006, 2002; Enderle and Wolfe, 1987). This system does not rely on muscle proprioceptors or real time visual feedback to ensure accuracy of movement because the eye movements occur too fast. Instead, a complex neural network involving the mesencephalon, cerebellum, brainstem and the cerebrum keeps track of the eye movement.

A typical experiment for recording saccades has the subject sitting before a horizontal target display of small light emitting diodes (LEDs) as shown in Fig. 1.3 (left). The subject is instructed to maintain their eyes on the lit LED by moving their eyes as fast as possible to avoid errors. A saccade is made by the subject when the active LED is switched off and another LED is switched on. Eye movements can be recorded using a variety of techniques, including electrooculography, video oculography, scleral search coil and infrared oculography (shown in Fig. 1.3 (right)). A typical saccade is shown in Fig. 1.4, with a latent period of approximately 100 ms and amplitude of 10° with duration of approximately 60 ms. Saccadic eye movements are conjugate and ballistic, with a typical duration of 30-100 ms and a latency of 100-300 ms. The latent period is thought to be the time interval during which the CNS determines whether to make a saccade and, if so, calculates the distance the eyeball is to be moved, transforming retinal error into transient muscle activity. Also shown in this figure is the velocity of the saccade with a peak velocity of approximately $400°\text{s}^{-1}$.

Generally, saccades are extremely variable, with wide variations in the latent period, time to peak velocity, peak velocity, and saccade duration. Furthermore, variability is well coordinated for saccades of the same size; saccades with lower peak velocity are matched with longer saccade durations, and saccades with higher peak velocity are matched with shorter saccade durations. Thus, saccades driven to the same destination usually have different trajectories.

To appreciate differences in saccade dynamics, it is often helpful to describe them with saccade main sequence diagrams (Bahill et al., 1975; Enderle and Wolfe, 1988; Harwood et al., 1999). The main sequence diagrams plot saccade peak velocity–saccade magnitude, saccade duration–saccade magnitude, and saccade latent period–saccade magnitude. The saccade size or amplitude is the angular displacement from the initial position to its destination. The size of a saccade ranges from less than a degree (microsaccades) to 45° in both the nasal (toward the nose) and temporal (toward the temple) directions. Peak or maximum velocity occurs at approximately half the duration of the saccade for small saccades less than 15° (Bahill et al., 1975). The duration of a saccadic eye movement

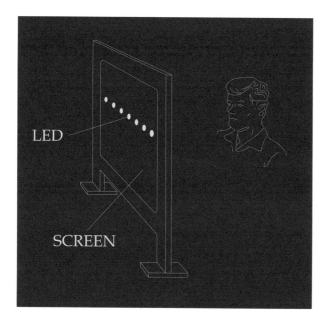

Figure 1.3: Experimental setup to record a saccade (left) and an infrared eye movement recorder (right). The eye movement recorder is based on the design by Engelken et al., 1984.

is the time from the start to the end of a saccade. Duration is usually hard to determine from the saccade amplitude vs. time graph, but it is more easily seen in the velocity vs. time graph as shown in Fig. 1.4. Saccade durations can range from approximately 30 ms for saccades less than 5°, and up to 100 ms for large saccades. For saccades greater than 7°, there is a linear relationship between saccade amplitude and duration. The latent period is the time interval from when a target appears until the eyes begin to move.

Shown in Fig. 1.5 are the main sequence characteristics for a subject executing 26 saccades. The subject actually executed 52 saccades in both the positive and negative directions with only the results of the saccades in the positive direction displayed in Fig. 1.5 for simplicity. Note that saccade characteristics moving to the left are different from those moving to the right. The solid lines in the figures include a fit to the data. Peak velocity-saccade magnitude is basically a linear function until approximately 15°, after which it levels off to a constant for larger saccades. Many researchers have fit this relationship to an exponential function. The line in graph (A) is fitted to the nonlinear equation

$$v_{\max} = \alpha \left(1 - e^{-\frac{x}{\beta}} \right) \tag{1.1}$$

where $v_{\max}$ is the maximum velocity, x the saccade size, and the constants α and β evaluated to minimize the summed error squared between the model and the data. Note that α is to represent the

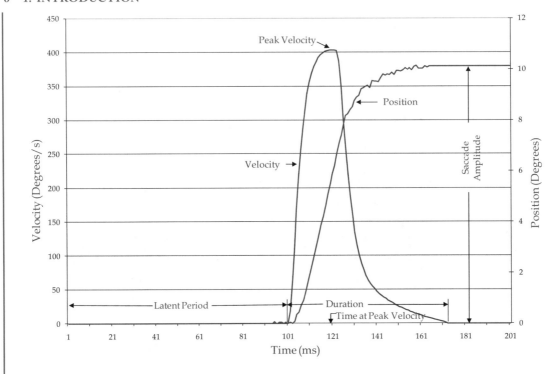

Figure 1.4: A 10° saccade with various indices labeled.

steady-state peak velocity-saccade magnitude curve and β is to represent the "time constant" for the peak velocity-saccade magnitude curve. For this data set for positive eye movements, α equals 825, and β equals 9.3. A similar pattern is observed with eye movements moving in the negative direction, but the parameters are $\alpha = 637$ and $\beta = 6.9$, which are typically different from the values computed for the positive direction. The exponential shape of the peak velocity–saccade amplitude relationship might suggest that the system is nonlinear if a step input to the system is assumed. A step input provides a linear peak velocity–saccade amplitude relationship. In fact, the saccade system is not driven by a step input but rather a more complex pulse-step waveform as discussed in Chapter 3. Thus, the saccade system cannot be assumed to be nonlinear solely based on the peak velocity-saccade amplitude relationship. The input to the saccade system is discussed more fully in Chapters 3 and 6.

Shown in Fig. 1.5B are data depicting a linear relationship between saccade duration–saccade magnitude. If a step input is assumed, then the dependence between saccade duration and saccade magnitude also might suggest that the system is nonlinear. A linear system with a step input always has a constant duration. Since the input is not characterized by a step waveform, the saccade system cannot be assumed to be nonlinear solely based on the saccade duration–saccade magnitude relationship.

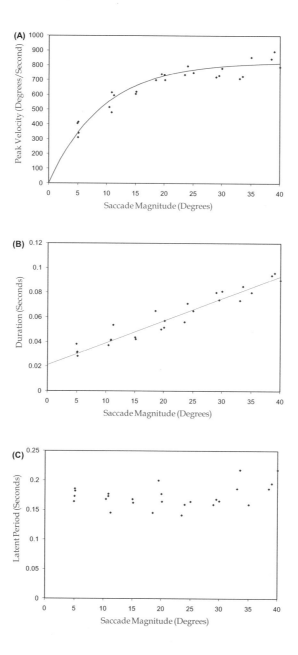

Figure 1.5: Main Sequence Diagrams for positive saccades. Similar shapes are observed for negative saccades. (A) Peak velocity-saccade magnitude, (B) saccade duration-saccade magnitude, and (C) latent period-saccade magnitude for 26 saccadic movements by a single subject. Adapted from: Enderle, J. (1988). Observations on pilot neurosensory control performance during saccadic eye movements. *Aviation, Space, and Environmental Medicine*, 59: 309.

Shown in Fig. 1.5C is the latent period-saccade magnitude data. It is quite clear that the latent period does not show any linear relationship with saccade size, i.e., the latent period's value appears independent of saccade size. However, some other investigators have proposed a linear relationship between the latent period and saccade magnitude. This feature is unimportant for the presentation in this book since in the development of the oculomotor plant models, the latent period is implicitly assumed within the model.

Because of the complexity of the eye movement system, attention is restricted to horizontal fast eye movements. In reality, the eyeball is capable of moving horizontally, vertically and torsionally. An appropriate model for this system would include a model for each muscle and a separate controller for each muscle pair. The development of the horizontal saccadic eye movement models in this book are historical and are presented in increasing complexity with models of muscle introduced out of sequence so that their importance is fully realized. Not every oculomotor model is discussed. A few are presented for illustrative purposes.

1.3   SMOOTH PURSUIT SYSTEM

Smooth pursuit is a voluntary eye movement that allows tracking of a slow moving target (with maximum velocity under $50 - 70°\text{s}^{-1}$) to maintain its position on the fovea. During smooth pursuit, vision remains clear, unlike a saccade where vision is interrupted. A time delay of approximately 100-200 ms occurs before smooth pursuit begins after acquiring a target to track. The smooth pursuit system responds to both position and velocity errors, but the velocity seems to be more important, especially in initiating the movement (Leigh and Zee, 1999). Oftentimes, smooth pursuit involves a saccade to bring the fovea on the target, followed by a continuous eye movement that matches the velocity of the target so that the target does not slip off the fovea. In general, smooth pursuit eye movements are initiated only when following a target (or the remembered motion of a target); that is, without visual stimulus, voluntary smooth pursuit eye movements do not occur. This is in contrast to a saccade, which can be voluntarily elicited. The performance of smooth pursuit is increased with predictable target movements such that the subject is able to perfectly track the motion of the target. To achieve such performance, a predictive mechanism is postulated to control the smooth pursuit system using an internal target velocity signal (Bahill and McDonald, 1983; Bahill and Harvey, 1986; Bahill and Hamm, 1989; Becker and Fuchs, 1985). Typically, humans tracking a periodic target quickly lock onto the target and track it with no latency, perhaps using an internal model of the target movement. The performance of the smooth pursuit system greatly varies among individuals, and it even varies for the same individual tested on different days. Smooth pursuit is affected by the quality stimulus, interest in following the stimulus, medications and age.

In the clinic or research laboratory, two types of stimuli are used to study the smooth pursuit system: a predictable sine-wave signal and a constant-velocity signal (ramp position input). A pseudo-random (sum of sine-waves) stimulus is also used to eliminate the predictability of the sine waveform. From the data collected, performance is evaluated using gain (peak eye velocity/peak target velocity) and phase (time offset of the output and input waveforms). Others use cross-spectral density analysis to evaluate central and peripheral pathology (Wolfe et al., 1978). Ideal performance for gain is close to 1.0 with a very small phase shift. Performance is usually best if the target moves slowly (< 1.0 Hz) and the amplitude of the movement is small (<±5°). If the amplitude is increased with frequency kept at 1 Hz, the smooth pursuit system does not operate as well. Smooth pursuit performance is better for predictable target motions than nonpredictable target motions.

As the target moves across visual space, the eyes track until they reach the limit of the field of vision, then the eyes move in the opposite direction with a saccade, acquiring a new target to track, with the process repeating itself. This type of eye movement pattern looks like saw-tooth time course and is called nystagmus. To keep up with the target in the real world, tracking often consists of smooth pursuit, interrupted by catch-up saccades.

1.4   VESTIBULAR OCULAR REFLEX EYE MOVEMENTS

The vestibular system holds the visual field steady during head rotations by rotating the eyes in the opposite direction to keep the target on the fovea. Vestibular ocular reflex (VOR) eye movements have a shorter latency (< 16 ms) than the saccadic or smooth pursuit eye movements because the labyrinth of the inner ear provides a signal to move the eyes faster than the visual system.

Within each ear, the labyrinth is divided into three parts: cochlea, the three semicircular canals, and endolymphatic sacs called the utricle and saccule that forms the otolith organs. The canals are oriented approximately perpendicular to each other, providing the system with the ability to sense angular rotation. Within the system are sensory cells with hairs that project directly into a gelatinous substance that bends when it flows past them during angular rotation. These sensors provide signals to the eye movement system during rotational head movements, which cause the eyes to move in the opposite direction the head is moving. Linear translational movements of the head stimulate the otolith organs. Normally in the human, angular rotation cause movements of the eyes – linear acceleration does not result in eye movements.

The function of VOR is to match the velocity of the eye to that of the head with a smooth pursuit movement to keep the target on the fovea. If the rotation continues past the visual field, the movement is interrupted by a saccade in the opposite direction, and then the smooth pursuit movement continues (nystagmus).

1.5   OPTOKINETIC EYE MOVEMENTS

Optokinetic eye movements occur when a large, full-field image moves uniformly across the retina during head rotations. The purpose of optokinetic eye movements is to help stabilize the retinal image during head rotation or translation. Similar to the nystagmus of VOR, after approximately a 1 to 2 s delay, the eyes move with the same velocity as the image, and then execute a saccade in the opposite direction when the full range of eye motion is complete. During the slow phase of the movement, the eyes are not tracking or attempting to keep the target on the fovea but simply stabilizing the image on the retina.

While both smooth pursuit and optokinetic eye stimulations could occur simultaneously when a small target is moving on a full-field image (providing contradictory responses), only the smooth pursuit response occurs. The smooth pursuit response is used to keep the target on the fovea, moving in the same direction as the target. The optokinetic eye movement response is just the opposite, keeping the image stabilized on the retina. In these situations, the smooth pursuit movement is the more important response and is executed rather than the optokinetic eye movement.

Optokinetic eye movements can be elicited by having the subject view a slowly rotating striped drum that fills the visual field (Carpenter, R., 1988). The eyes follow the strip on the drum with a smooth pursuit movement until the gaze moves to the end of the visual field, after which a saccade occurs in the opposite direction. After the saccade, the eyes acquire another strip on the drum and follow it, continuing this process of smooth pursuit followed by a saccade. This type of eye movement is called optokinetic nystagmus.

1.6   VERGENCE

All of the previous eye movements presented are conjugate eye movements, with each eye driven by the same neural control. Vergence eye movements allow us to move our eyes under separate controls when the target moves closer or farther away. When viewing a target at a far distance, the line of sight for each of the eyes to the target is straight ahead, approximately parallel with each other. As the target comes closer, the eyes change their direction by moving nasally, using a different neural controller for each eye.

CHAPTER 2

Smooth Pursuit Models

2.1 INTRODUCTION

The first model of the smooth pursuit system was published by Young and Stark in 1963, and it involves a sampled-data visual system with negative feedback. The input to this model is the target position and the output is eye position. The input to the forward-path is the retinal error position *(REP)*, the error between the fovea and target. The forward-path includes:

- Sampled-data element

- Two limiters to restrict abrupt changes and velocity exceeding 30°

- Overdamped second-order oculomotor plant

- Two integrators, one to translate position into velocity and the other to translate velocity into position.

While this model is able to describe some features of the smooth pursuit system, it does not accurately depict many aspects of the real system.

Many models have been proposed since 1963 that are increasingly more complex and able to more accurately depict additional aspects of the smooth pursuit system. Key components of these models are that the stimulus to the smooth pursuit system is primarily the target's velocity and a transport delay in the forward-path. The brain uses the difference between the target's velocity and the eye velocity, called retinal error velocity (*REV*), to move the eye so that it eventually moves at the same velocity as the target. The transport delay represents the time interval starting when the target appears on the retina, the time it takes for the signal to be sent from the retina to the brain sites responsible for the smooth pursuit, signal analysis by the brain, and the time it takes to send the command signal to the muscles of the eye that results in an eye movement. Some newer models include both the retinal error velocity and retinal error position to accurately model the response during the start of the stimulus and the response after the stimulus had ended. Before discussing some of these models, we will introduce a simple system to illustrate several important points.

2.2 A SIMPLE MODEL OF THE SMOOTH PURSUIT SYSTEM

Consider the smooth pursuit model for horizontal eye movements shown in Fig. 2.1. The input to the model is the target velocity, $\dot{T}$, and the output is eye velocity, $\dot{E}$. In the forward-path is the transport delay, TD, the controller, G_C, and the oculomotor plant, G_P. The feedback element is unity. The retinal error velocity is given by $REV = \dot{T} - \dot{E}$.

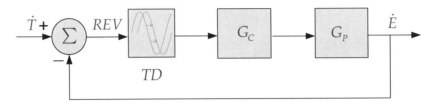

Figure 2.1: Simple smooth pursuit feedback model. All functions and variables are functions of the Laplace variable s.

2.2.1   A SIMPLE MODEL WITHOUT A TIME DELAY

Initially, we consider the case when the time delay is zero. Here, the system's closed-loop transfer function, G_{CL}, is given by

$$G_{CL} = \frac{\dot{E}}{\dot{T}} = \frac{G_C G_P}{1 + G_C G_P}.$$ (2.1)

When a retinal velocity error is present, the error driving the system is $\dot{T}$. As the eyes move toward the target, the error becomes progressively less until it reaches a minimum value. The retinal velocity error is given by

$$REV = \frac{1}{1 + G_C G_P} \dot{T}.$$ (2.2)

For the eyes to be traveling at almost the same speed as the target, the magnitude of $1 + G_C G_P$ must be very large over the appropriate range of s. Once this happens, the REV is very small, resulting in a very small input to the forward-path necessary to keep the eyes traveling at almost the same speed as the target. Another way to think of this is using Eq. (2.1), with $G_C G_P \gg 1$, then $\dot{E} \approx \dot{T}$. We will illustrate this point by examining the REV as G_C increases.

UNDERDAMPED OCULOMOTOR PLANT

In the following cases, we assume that $G_C = K$, $G_P = \frac{\omega_n^2}{s^2 + 2\zeta\omega_n s + \omega_n^2}$ and $\dot{T} = \mathcal{L}\{3\cos(1.885t)\}$. The oculomotor plant, G_P, is based on a 1954 second-order model by Westheimer, with $\zeta = 0.7$ and $w_n = 120$ as described in Section 3.2.1. Note that the oculomotor plant in this system is underdamped, and oscillatory behavior is evident during the transient response. The input is a slow sinusoid with frequency 0.3 Hz and amplitude of 3. Substituting G_C and G_P into Eq. (2.1), we have

$$G_{CL} = \frac{\frac{K\omega_n^2}{s^2 + 2\zeta\omega_n s + \omega_n^2}}{1 + \frac{K\omega_n^2}{s^2 + 2\zeta\omega_n s + \omega_n^2}} = \frac{K\omega_n^2}{s^2 + 2\zeta\omega_n s + (K+1)\omega_n^2}.$$ (2.3)

We assume zero initial conditions for the eye. The form of the solution for this system is given by

$$\dot{e}(t) = e^{\alpha t} \left(B_1 \cos(\beta t) + B_2 \sin(\beta t) \right) \\ + B_3 \cos(1.885t) + B_4 \sin(1.885t) \tag{2.4}$$

where $\dot{e}(t)$ is the velocity of the eye, $\alpha \pm j\beta$ are the roots of the characteristic equation, and B_i are the constants evaluated from the zero initial conditions and the input.

Consider the case where $K = 2$. The closed-loop transfer function is given by

$$G_{CL} = \frac{28,800}{s^2 + 168s + 43,200} . \tag{2.5}$$

The roots of the characteristic equation are $-84 \pm j190.12$ and the solution given by

$$\dot{e}(t) = e^{-84t} \left(-2 \cos(190.12t) - 0.8835 \sin(190.12t) \right) \\ + 2 \cos(1.885t) + 0.0147 \sin(1.885t) . \tag{2.6}$$

The MatLab SIMULINK response for this case is shown in Fig. 2.2. The transient response is quite

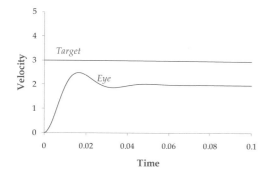
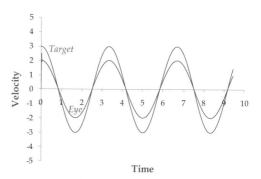

Figure 2.2: (Left) Smooth pursuit response for an underdamped oculomotor plant with gain $K = 2$. Response to steady state illustrating the transient response due to the oculomotor plant. (Right) Steady state response with scale that hides the transient response.

brief, dying out in approximately 0.06 s as shown in Fig. 2.2 (Left). The *REV* is approximately 1 after steady-state is reached as shown in Fig. 2.2 (Right).

Next, we increase G_C to 20, with results shown in Fig. 2.3. The nature of the response changes according to Eq. (2.3). The roots of the characteristic equation are $-84 \pm j534.46$, and $B_3 = 2.8572$. The time to steady-state stays at approximately 0.06 s. The number of oscillations in the transient response have increased since the imaginary component of the root increased. The *REV* is 0.1428 (i.e., $3.0 - 2.8572$) after steady-state is reached as shown in Fig. 2.3 (Right).

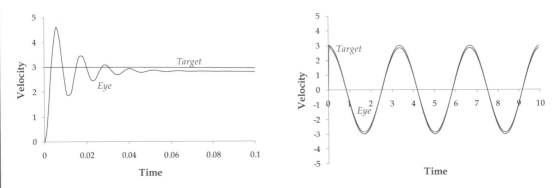

Figure 2.3: (Left) Smooth pursuit response for an underdamped oculomotor plant with gain $K = 20$. Response to steady state illustrating the transient response of the oculomotor plant. (Right) Steady state response with scale that hides the transient response.

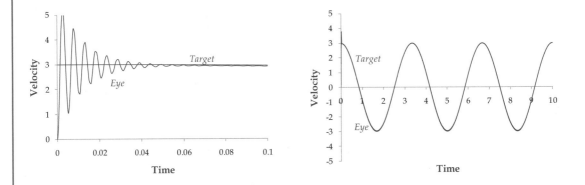

Figure 2.4: (Left) Smooth pursuit response for an underdamped oculomotor plant with gain $K = 100$. Response to steady state illustrating the transient response of the oculomotor plant. (Right) Steady state response with scale that hides the transient response. The target and eye velocity almost completely overlap except for $0 \leq t < 0.06$.

Next, we increase G_C to 100, with results shown in Fig. 2.4. The roots of the characteristic equation are $-84 \pm j1203.1$, and $B_3 = 2.9703$. The time to steady-state stays at approximately 0.06 s. As before, the number of oscillations have increased since the imaginary component of the root increased. The REV is 0.0297 after steady-state is reached as shown in Fig. 2.4 (Right).

As demonstrated, the negative feedback system described here offers excellent performance with small error signals. The sensitivity to parameter variations in $G_C G_P$ is quite small, especially when $G_C G_P \gg 1$.

Negative feedback is rather typical in a physiological system as it provides a much faster response than an open-loop system, it is very accurate, and it is relatively insensitive to parameter changes in the forward path. For instance, consider changes in G_C that might occur with aging. Here, the overall system performance is minimally impacted even with a 50% change; that is, if at birth $G_C = 10$ and then falls to 5 with aging, the closed-loop gain goes from 0.91 to 0.83, a rather small change in the overall system given the large change in the gain.

Feedback control systems are sensitive to changes in feedback. Suppose that the feedback element in Eq. (2.1) is given by H rather than 1. The closed-loop transfer function is

$$G_{CL} = \frac{\dot{E}}{\dot{T}} = \frac{G_C G_P}{1 + H G_C G_P} = \frac{1}{\frac{1}{G_C G_P} + H} . \tag{2.7}$$

If $G_C G_P \gg 1$, then $G_{CL} \approx \frac{1}{H}$. Thus, we have $\dot{E} \approx \frac{\dot{T}}{H}$. Any changes in H are directly evident in the system response.

OVERDAMPED OCULOMOTOR PLANT

Before moving into the case involving a time delay, we wish to consider an overdamped oculomotor plant with roots at $s_1 = 5 = \frac{1}{0.2}$ and $s_2 = 142.8571 = \frac{1}{0.007}$ (pole locations are from Robinson, D., 1973) to investigate the oscillations in the response, where

$$G_P = \frac{714.2857}{(s + 5)(s + 142.8571)} = \frac{1}{(0.2s + 1)(0.007s + 1)} \tag{2.8}$$

and

$$G_{CL} = \frac{G_C G_P}{1 + G_C G_P} = \frac{\frac{714.2857K}{(s+5)(s+142.8571)}}{1 + \frac{714.2857K}{(s+5)(s+142.8571)}}$$

$$= \frac{714.2857K}{s^2 + 147.8571s + 714.2857(K + 1)} . \tag{2.9}$$

With these parameters, the eye velocity solution has the form

$$\dot{e}(t) = B_1 e^{s_1 t} + B_2 e^{s_2 t} + B_3 \cos(1.885t) + B_4 \sin(1.885t) \tag{2.10}$$

for values of K less than 6.7, and

$$\dot{e}(t) = e^{\alpha t}(B_1 \cos(\beta t) + B_2 \sin(\beta t)) \\ + B_3 \cos(1.885t) + B_4 \sin(1.885t) \tag{2.11}$$

for values of K greater than 6.7.

With $G_C = 2$, the natural response is shown in Fig. 2.5 (Left), an overdamped response. The roots of the characteristic equation are -131.57 and -16.2872, and the time to steady-state is approximately 0.3 s. Notice that there is a sizeable REV as shown in Fig. 2.5 (Right).

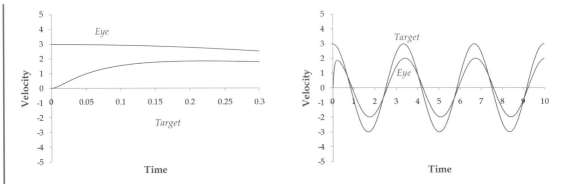

Figure 2.5: (Left) Smooth pursuit response for overdamped oculomotor plant with gain $K = 2$. Response to steady state illustrating the transient response due to the oculomotor plant. (Right) Steady state response with scale that hides the transient response.

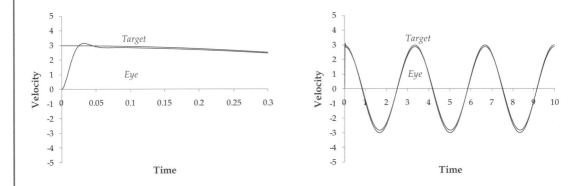

Figure 2.6: (Left) Smooth pursuit response for overdamped oculomotor plant with gain $K = 20$. Response to steady state illustrating the transient response due to the oculomotor plant. (Right) Steady state response with scale that hides the transient response.

With $G_C = 20$, the natural response is shown in Fig. 2.6 (Left). The roots of the characteristic equation are $-73.92 \pm j97.64$, which has an underdamped natural response. The *REV* is quite close to zero in this case.

As G_C increases, the real part of the root stays at -73.92, and the imaginary part increases. Thus, there is an increase in the number of oscillations in the natural response as G_C increases and the *REV* decreases. To have an *REV* approximately the same value as the underdamped case requires $K \approx 100$, giving us a closed-loop response that is underdamped, even though the oculomotor plant is overdamped.

OVERDAMPED OCULOMOTOR PLANT WITH EYE AND VELOCITY INPUTS

Some models of the smooth pursuit system include retinal error position and velocity in the forward path as shown in Fig. 2.7. These models typically use the same overdamped oculomotor plant

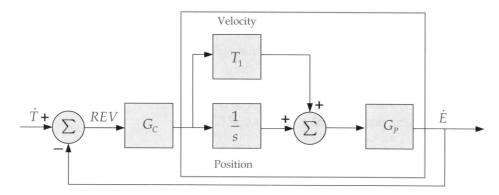

Figure 2.7: Simple smooth pursuit feedback model with retinal error position and velocity in the forward path. $T_1 = 0.2$.

described before as

$$G_P = \frac{1}{(sT_1 + 1)(sT_2 + 1)} = \frac{1}{(0.2s + 1)(0.007s + 1)} = \frac{714.2857}{(s + 5)(s + 142.8571)} . \quad (2.12)$$

Note that the velocity block, T_1, in Fig. 2.7 has the same value of one of the roots, $s_1 = 5 = \frac{1}{T_1}$. Now, we wish to replace the items within the blue box in Fig. 2.7 with an equivalent representation.

$$\dot{E} = \frac{\left(T_1 + \frac{1}{s}\right)}{(sT_1 + 1)(sT_2 + 1)} (REV \times G_C)$$

$$= \frac{(sT_1 + 1)}{s} \frac{1}{(sT_1 + 1)(sT_2 + 1)} (REV \times G_C)$$

$$= \frac{1}{s(sT_2 + 1)} (REV \times G_C) . \quad (2.13)$$

Since $T_2 = 0.007$, $(sT_2 + 1) \approx 1$, and

$$\dot{E} \approx \left(\frac{1}{s}\right) (REV \times G_C) . \quad (2.14)$$

It should be clear from Eq. (2.14) that eye velocity, $\dot{E}$, approximately equals the integral of the $REV \times G_C$.

Suppose we replace the items in the blue box in Fig. 2.7 with an integrator as in Eq. (2.14). The closed-loop transfer function is:

$$G_C = \frac{\frac{G_c}{s}}{1 + \frac{G_c}{s}} = \frac{G_C}{s + G_C} .$$ (2.15)

Here the root for the closed-loop system is $-G_C = -K$. Thus, as K increases to reduce *REV*, the time constant, $\frac{1}{K}$, decreases, as does the time for the natural response to reach steady state. Shown in Figures 2.8 and 2.9 are simulations with $K = 20$ and 100. As illustrated, the time for the natural response to reach steady-state is much smaller for $K = 100$ than $K = 20$. Both simulations track the target very well after this time with a very small *REV*.

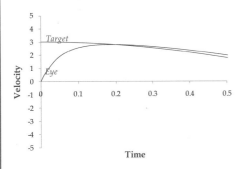
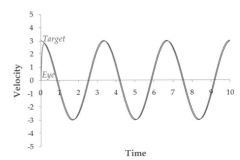

Figure 2.8: (Left) Smooth pursuit response for an integrator oculomotor plant with gain $K = 20$. Response to steady state illustrating the transient response due to the oculomotor plant. (Right) Steady state response with scale that hides the transient response.

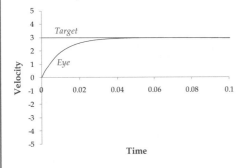
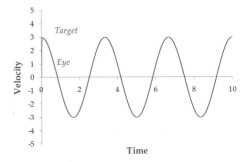

Figure 2.9: (Left) Smooth pursuit response for an integrator oculomotor plant with gain $K = 100$. Response to steady state illustrating the transient response due to the oculomotor plant. (Right) Steady state response with scale that hides the transient response.

The problem with the model in Fig. 2.7 is that it requires that the gain multiplying the velocity equal T_1, a totally unreasonable assumption. Furthermore, the oculomotor plants used in this section do not realistically describe the actual muscle and eye ball system as will be demonstrated in the next chapters on saccades. However, the actual oculomotor plant is not critical in modeling the smooth pursuit system since the input is very slow as compared with the time it takes the natural response to reach steady state. As we will see in Section 2.3.1, no loss in accuracy is observed by setting $G_P = 1$.

2.2.2   A SIMPLE MODEL WITH A TRANSPORT DELAY

Now, we consider the case when a transport delay, or time delay is added to the system. Here, the system's closed-loop transfer function, G_{CL}, is given by

$$G_{CL} = \frac{\dot{E}}{\dot{T}} = \frac{G_C G_P e^{-sTD}}{1 + G_C G_P e^{-sTD}} .$$ (2.16)

The time delay for the smooth pursuit system is 100-200 ms. The time delay introduces a phase lag into the system without changing the magnitude of the closed-loop transfer function (no poles or zeros are added). The addition of a time delay to a system can affect the stability of the system because of the additional phase lag. To keep the system stable with a time delay and its commensurate phase lag, the gain, G_C, often needs to be reduced from values used in the previous section. As shown before, reduction of the gain increases the *REV* of the response.

Consider the system shown in Fig. 2.1 with an underdamped oculomotor plant, $TD = 0.007$ s and $G_C = K = 2$. The response is shown in Fig. 2.10, which has large oscillations that occur in the

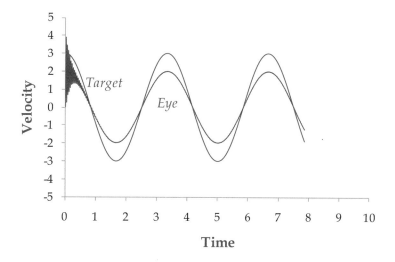

Figure 2.10: Smooth pursuit response for the system with $TD = 0.007$ s and $K = 2$.

beginning of the movement, but then it declines with time. If $TD = 0.01$ s with $K = 2$, the system is unstable. The change in stability can also be appreciated by plotting a Bode or Nyquist plot.

With $TD = 0.1$ s as in the real smooth pursuit system, the largest value of K that results in a stable system is $K = 1$. The simulation for this case is shown in Fig. 2.11.

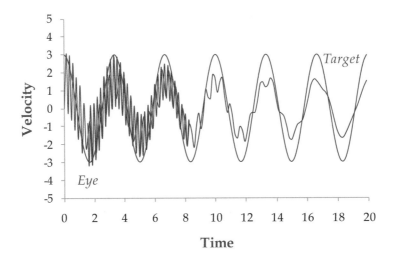

Figure 2.11: Smooth pursuit response for the system with $TD = 0.1$ s and $K = 1$.

As demonstrated, a system with a time delay and a 2nd order oculomotor plant cannot have a *REV* close to zero since the large gain causes the negative feedback system to become unstable. These features are present in the simple model presented here, which produce growing oscillations unlike those seen in the data.

Suppose we replace $G_C G_P$ with a gain element and integrator—the elements that result when a position and velocity error are used with a 2nd order oculomotor plant. In this case, the smooth pursuit system tracks the target quite well with a gain of 10 and a time delay of 0.1 s as shown in Fig. 2.12. In Fig. 2.12 (Left), oscillations are observed in the natural response. In Fig. 2.12 (Right), the tracking at steady-state still has the time delay between the input and the response, but the magnitude is approximately the same. If the gain is increased to 20, the system is unstable.

A normal human is able to track with an overall gain close to 1 with no time delay following predictable targets such as the sine wave used here. Humans are able to accurately track within the first quarter of the initial sine wave target motion. Small damped oscillations of 2-4 Hz are often observed at the beginning of the movement. Nonpredictable target motions are not tracked by humans as accurately as predictable motions. Models of the form in Figures 2.1 and 2.7 cannot account for normal smooth pursuit tracking with no time delay. As demonstrated next, there needs to be a predictive property included in the model for it to behave like the real system.

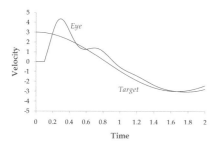
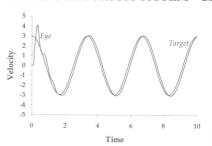

Figure 2.12: (Left) Smooth pursuit response with gain $K = 10$. Response to steady state illustrating the transient response due to the oculomotor plant. (Right) Steady state response.

2.3 MORE COMPLEX MODELS OF THE SMOOTH PURSUIT SYSTEM

Now, we will consider more complex models of the smooth pursuit system beginning with the model shown in Fig. 2.13, loosely based on the one by Yasui and Young, 1975. The major difference

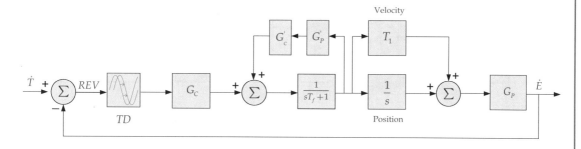

Figure 2.13: Smooth pursuit model loosely based on the one by Yasui and Young, 1975.

between this model and the previous ones is the inclusion of an internal positive feedback loop, called the efference loop, with $G_C^1 G_P^1$. With the efference loop, the input to the system has been modified from the *REV* to an internal representation within the brain that represents the motion of the target. The element $\frac{1}{sT_f+1}$ is a filter. The thought is that the efference loop cancels the outer visual feedback loop, and, therefore, the system operates in an open-loop mode and can operate with higher gains. While this model is stable for larger gains than those in the previous section, it does not eliminate the time delay in the response.

2.3.1 BAHILL MODEL

Bahill and coworkers introduced a target-selective adaptive control model of the human smooth pursuit and saccadic eye movement system that produces accurate, zero-latency smooth pursuit target tracking as observed in the data (Bahill and Hamm, 1989; Bahill and Harvey, 1986; Bahill and McDonald, 1983). The overall model was a major step forward from previous models since it includes a saccade branch, a predictor to reduce the time delay (called the target-selective adaptive controller), and a smooth pursuit branch. This model is able to reproduce eye tracking data that typically has saccades and smooth pursuit movements. Saccades usually occur at the beginning of the movement, allowing the eyes to catch up with the target, and then the brain tracks the target with the smooth pursuit system. Four models for the smooth pursuit system were considered: integrator $\left(\frac{K}{s}\right)$, a leaky integrator $\left(\frac{K}{\tau s+1}\right)$, a critically damped 2nd order system, and an overdamped 2nd order system. After analysis, the leaky integrator was selected as the best representative for the smooth pursuit system.

The model for the overall system is shown in Fig. 2.14. There are three major components in

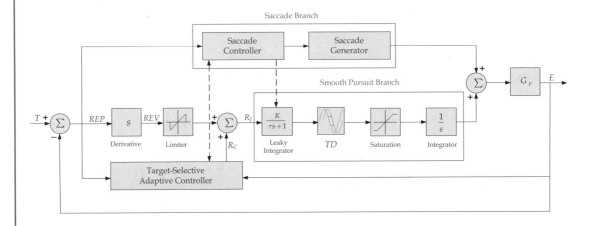

Figure 2.14: Target-selective adaptive control model primarily based on Bahill and coworkers. The input is target position and the output is eye position. The smooth pursuit and saccade branches are enclosed in the blue boxes.

the eye movement system: smooth pursuit system, saccade system and the target-selective adaptive controller. The overall system operates with negative unity feedback. The input to the overall system is target position and the output is eye position. The saccade branch corrects position errors and the smooth pursuit branch corrects velocity errors, helped by the target-selective adaptive controller that reduces time delay errors.

The original oculomotor plant, G_P, was based on Bahill and coworkers 6th order linear homeomorphic model (Bahill et al., 1980). This model produced accurate saccades, but it was judged more complex than needed for smooth pursuit movements. They then adopted the 2nd order underdamped oculomotor plant by Westheimer, which provided sufficiently accurate eye movements. The 6th order G_P required a pulse-step saccade generator output, and the 2nd order G_P required a step saccade generator output. Here, we use an oculomotor plant, G_P, set to 1 for simplicity and without any loss in accuracy. The output of the saccade generator is the eye position error.

Saccade Branch   The saccade branch consists of two elements: a saccade controller and a saccade generator. The saccade controller monitors the error between target and eye position with a time delay of 150 ms. When the target-selective adaptive controller is operational and the error exceeds the threshold set at 0.5°, the saccade controller sends a command to the saccade generator to execute a saccade. In addition, when the target-selective adaptive controller is turned off and the error exceeds the threshold set at 0.3°, the saccade controller sends a command to the saccade generator to execute a saccade.

Smooth Pursuit Branch   The smooth pursuit system includes:

- One input that consists of a derivative to transform *REP* into *REV*, which is then limited to eliminate tracking target velocities greater than $70°\text{s}^{-1}$

- A second input which is the output of the target-selective adaptive controller, R_C

- A leaky integrator with a time constant of τ

- A time delay

- A saturation element to eliminate eye velocities greater than $60°\text{s}^{-1}$

- An integrator to transform the velocity signal into a position signal

The input to the smooth pursuit system is the perceived velocity, R_J, which consists of the limited *REV* and the output of the target-selective adaptive controller, R_C. R_C allows the smooth pursuit system to predict future target velocities from predictable target movements and to adjust for the dynamics of the entire system. The target-selective adaptive controller has inputs of the target position and the eye position.

Target-Selective Adaptive Controller   The overall function of the target-selective adaptive controller is to monitor position and velocity errors, and when detected, provide a signal that allows the eyes to track the target without any time delay. Various thresholds are active in this unit. It only operates when the target moves smoothly, that is, when there are no discontinuities, and between frequencies 0.1 and 1 Hz. Frequencies below 0.1 Hz are considered stationary. When the target is stationary for more than 50 ms and the position error is exceeds 0.3°, or when the velocity error exceeds $3°s^{-1}$, the output of the target-selective adaptive controller is zero. When the target-selective adaptive controller is activated, it takes approximately one-quarter cycle of the target motion before it is able to track with no time delay or velocity error.

The target-selective adaptive controller is able to track sinusoids, parabolic waveforms (i.e., one with constantly increasing velocities), and cubic waveforms by menu selection. The controller is also able to track using a 2nd order difference equation, however, with less accuracy as demonstrated by Bahill and workers (not presented here).

The target-selective adaptive controller is able to interact with the two other branches as indicated in Fig. 2.14 with the dashed lines. The saccade controller inhibits the target-selective adaptive controller during a saccade, as well as the smooth pursuit branch. The target-selective adaptive controller reduces the threshold of the saccade controller when the position error is less than $0.3°$ and when the velocity error is less than $3°s^{-1}$.

The output of the target-selective adaptive controller, R_c, is created using an internal model of the target motion to predict future movements and eliminate the time delays between the target movement and the eye movement.

To be able to predict the target behavior, the dynamics of the system must be incorporated into R_c. Thus, when added to the target movement, the system is able to track without any delay. That is, if $\dot{T}(t)$ is the current target velocity, then R_c must produce $\dot{T}(t + TD)$, where TD is the time delay of the smooth pursuit system. The adaptive controller must also modify the previous R_c to compensate for the system dynamics. For the leaky integrator used here, $\left(\frac{K}{\tau s+1}\right)$, R_c is modified as

$$R_c(t) = \frac{1}{K}\left[\frac{d}{dt}\tau \dot{T}(t + TD) + \dot{T}(t + TD)\right] \tag{2.17}$$

that is, multiplying $\dot{T}(s)$ by the inverse of the leaky integrator, $\frac{(\tau s+1)}{Ke^{-TDs}}$, to compensate for the system dynamics.

As described previously, the target-selective adaptive controller is able to follow three different waveforms. Here, we will only consider a target moving sinusoidally, $\dot{T}(t) = A\omega \cos(\omega t)$ (i.e., target position is $A\sin(\omega t)$). This target motion is substituted into Eq. (2.17) to give

$$R_c(t) = \frac{A\omega}{K}\left[\cos\left(\omega\left(t + TD\right)\right) - \tau\omega\sin\left(\omega\left(t + TD\right)\right)\right] . \tag{2.18}$$

The target-selective adaptive controller estimates A, ω, K, TD and τ through a neural network in the brain.

OPEN-LOOP EXPERIMENTS

The smooth pursuit system is a closed-loop system given by Eq. (2.1). If the open-loop gain, G_C, is quite large, then the closed-loop is approximately 1. Negative feedback tends to hide the forward path making it is difficult to study its dynamics, i.e., $G_C G_P$. By opening the feedback loop, as in Fig. 2.15, one can study the internal workings of the system to more thoroughly investigate $G_C G_P$ without feedback. The open-loop transfer function is $G_{OL} = \frac{\dot{E}}{\dot{T}} = e^{-TDs}G_C G_P$. Keep in mind that when a loop is opened, the system no longer behaves normally and unusual behavior can result; for instance, if the forward path included an integrator and the input was a step, the system would respond with a ramp until it saturated.

Ideally, a loop can be opened by cutting the connection to the summer as depicted in Fig. 2.15. In animal studies, opening the loop can be carried out by lesion or freezing the tissue. In humans,

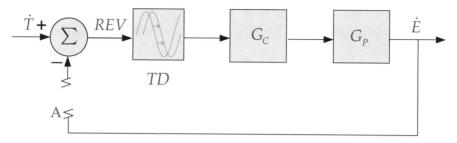

Figure 2.15: A closed-loop smooth pursuit system with the feedback loop opened at point A.

we typically trick the brain into seeing the system as opened. One simple technique for opening the loop for the smooth pursuit system is to have a subject tracking a sinusoidally moving target by increasing the target movement by the amount of movement by the eyes. This can be done with electronics controlling the target movement via software and using the recorded eye movement position. Thus, the eyes never catch up to the target, and, artificially, we have opened the loop. Bahill and Hamm identified some issues with this approach from subject to subject, as some subjects turned off their saccade system, others turned off the target-selective adaptive controller, some turned off both, and others kept both in place, giving rise to unevenness among reports in the literature (Bahill and Hamm, 1989).

Bahill and coworkers were able to focus on just the smooth pursuit branch by careful experimental design. By keeping the head stationary, the vestibulo-ocular movement was eliminated, and by keeping the target a fixed distance from the subject, the vergence movement was eliminated. They eliminated the target-selective adaptive controller by using unpredictable target movements and analyzing the first few seconds of the movement. A unique target waveform was used to eliminate saccades, and the speed of the target movement eliminated the limiter and the saturation element.

DATA, PARAMETER ESTIMATION AND THE SMOOTH PURSUIT MODEL

Data were collected on the time delays for the saccadic branch and smooth pursuit branch, eye and target velocity and the rise-time for the open- and closed-loop experiments. From this analysis, a final smooth pursuit model was determined.

To collect data for the saccade branch, the target movement consists of a series of step displacements, which did not elicit a smooth pursuit movement. From these movements, the time delay for the saccade branch was calculated as approximately 200 ms.

The time delay for smooth pursuit branch was calculated using sine waves and step-ramps, while keeping the loop closed. The transport delay for the sine waves was 176 ms and for the step-ramps 171 ms.

To determine the gain of the controller, G_c, the ratio of $\frac{\dot{E}}{\dot{T}}$ was calculated from the open-loop experiment with sine waves and ramps as the input. The gains obtained averaged 3.11 for the sine wave inputs and 1.61 for the ramp inputs. It was felt that the gain from the sine wave input was less reliable because of quick adaptation by the subjects. The closed-loop gain for the step-ramp input averaged 2.35.

Rise-Time Estimates   Next, the rise-time is used to estimate the time constant in the leaky integrator in both the closed-loop and open-loop experiments. In this analysis, we ignore the time delays since they are easily removed and simplifies the analysis. Thus, the forward path is $G_C G_P = \frac{K}{\tau s + 1}$, since $G_P = 1$.

The closed-loop transfer function is

$$\frac{\dot{E}}{\dot{T}} = \frac{G_C G_P}{1 + G_C G_P} = \frac{\frac{K}{\tau s + 1}}{1 + \frac{K}{\tau s + 1}} = \frac{K}{\tau s + 1 + K} \, . \tag{2.19}$$

With a step velocity input (a ramp displacement), $\dot{T} = \frac{1}{s}$, Eq. (2.19) is now written as

$$\dot{E} = \frac{K}{s \, (\tau s + 1 + K)} \tag{2.20}$$

which has a solution

$$\dot{e}(t) = \frac{K}{K + 1} \left(1 - e^{-\frac{(K+1)t}{\tau}} \right) = \frac{K}{K + 1} \left(1 - e^{-\frac{t}{\tau_{CL}}} \right) \, . \tag{2.21}$$

The ratio of steady-state eye velocity to target velocity from human data equals 0.67, which, from Eq. (2.21) gives $\frac{K}{K+1} = 0.67$ and $K = 2.3$. The rise-time, T_{RT}, is a function of the time constant of the system, τ_{CL}. Note that the rise-time is defined as the time it takes to go from 10% to 90% of steady state. Steady-state is $\frac{K}{K+1} = 0.67$. In general, the time it takes to get to 0.9 of steady state, $t_{0.9}$, is $2.3\tau_{CL}$, that is

$$0.9 \frac{K}{K + 1} = \frac{K}{K + 1} \left(1 - e^{-\frac{t_{0.9}}{\tau_{CL}}} \right) \tag{2.22}$$

which simplifies to $\ln(0.1) = -\frac{t_{0.9}}{\tau_{CL}}$, or $t_{0.9} = 2.3\tau_{CL} = 2.3 \frac{\tau}{K+1}$. Next, we find the time it takes to get to 0.1 of steady state, which can be shown equal to $0.1\tau_{CL}$. Thus, the rise-time is

$$T_{RT} = 2.2\tau_{CL} = 2.2 \frac{\tau}{K + 1} \, . \tag{2.23}$$

With $T_{RT} = 0.096$ and $K = 2.3$, we have $\tau = \frac{(1+K)T_{RT}}{2.2} = 142$ ms.

The open-loop transfer function is

$$\frac{\dot{E}}{\dot{T}} = \frac{K}{\tau s + 1} \, . \tag{2.24}$$

With a step velocity input, $\dot{T} = \frac{1}{s}$, and Eq. (2.24) is now

$$\dot{E} = \frac{K}{s \, (\tau s + 1)} \tag{2.25}$$

which has a solution

$$\dot{e}(t) = K \left(1 - e^{-\frac{t}{\tau}} \right). \tag{2.26}$$

The ratio of steady-state eye velocity to target velocity from human data equals 1.61. The rise-time here equals $T_{RT} = 2.2\tau$, or $\tau = \frac{T_{RT}}{2.2} = \frac{0.182}{2.2} = 83$ ms.

The final smooth pursuit model based on the previous results is given as

$$G_C = \frac{2e^{-0.15s}}{0.13s + 1}. \tag{2.27}$$

The saccade branch includes a time delay of 0.15 s, a sample and hold element set at 0.15 s, and a threshold detector to prevent a saccade occurring if the retinal error is less than 0.5 degrees.

Shown in Fig. 2.16 is the Simulink model of the target-selective adaptive controller for smooth pursuit and saccadic eye movements. The target position input is $3\sin(1.885t)$, and parameters are given as previously described. Shown in Fig. 2.17 are the position and velocity response to a sinusoidally moving target. Notice that it takes less than a period for the system to adequately track the target with virtually zero error. A few saccades occur early in the movement as seen in Fig. 2.17 (Top). Without the target-selective adaptive controller, the system is no longer able to track with zero latency, and saccades occur throughout the simulation.

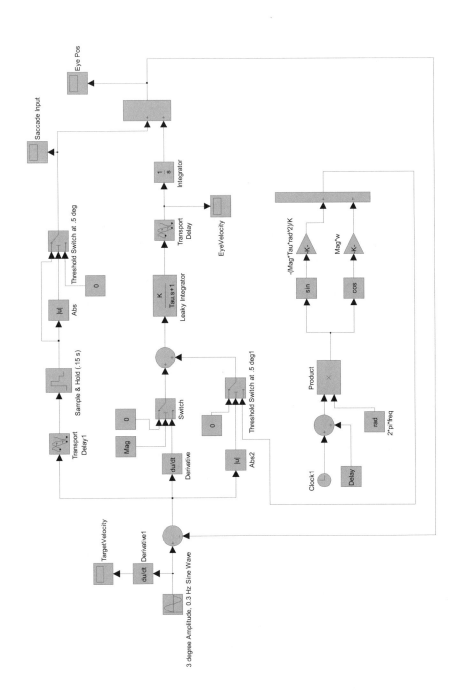

Figure 2.16: Simulink model of the target–selective adaptive controller for eye movements.

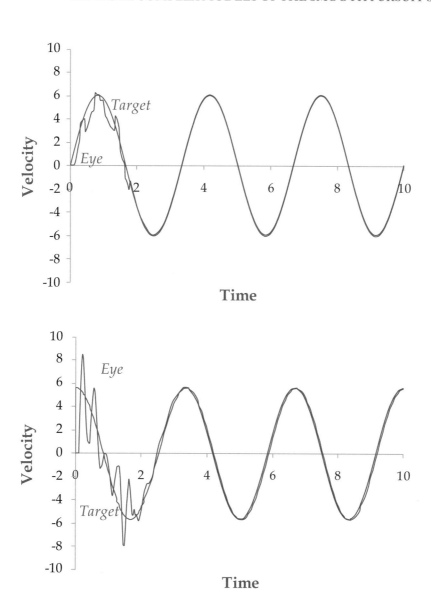

Figure 2.17: Graphs of the smooth pursuit response to a sinusoidal target obtained from Simulink simulations for the target-selective adaptive controller model of the smooth pursuit system. (Top) Smooth pursuit position response (red line) to a sinusoidally moving target (blue line). (Bottom) Smooth pursuit velocity response (red line) to a sinusoidally moving target (blue line).

CHAPTER 3

Early Models of the Horizontal Saccadic Eye Movement System

3.1 INTRODUCTION

The oculomotor plant and saccade generator are the basic elements of the saccadic system. The oculomotor plant consists of three muscle pairs and the eyeball. These three muscle pairs contract and lengthen to move the eye in horizontal, vertical, and torsional directions. Each pair of muscles acts in an antagonistic fashion due to reciprocal innervation by the saccade generator. For simplicity, the models described here involve only horizontal eye movements and one pair of muscles, the lateral and medial rectus muscle. We call these muscles antagonistic pairs because their activity opposes each other, a principle known as reciprocal innervation as described originally by Descartes, R. (1664). Most of the advances seen in models of the saccadic eye movement system involve more complete models of the oculomotor muscles and the neural network controlling these movements.

Four types of models are typically used to describe muscles: nonlinear models that incorporate the fundamental relationships of Hill, cross-bridge models, anatomical models, and linear and piece-wise linear models. Of the four, the linear muscle models prove to be the most popular in systems applications because of their relative mathematical ease in analysis and simplicity. In the past, linear muscle models, and linear oculomotor muscle models in particular, have been criticized for their failure to successfully account for the nonlinear interpretations of experimental evidence. The other model types are generally not used in oculomotor systems applications because of their complexity, except for the nonlinear homeomorphic model by Hsu and coworkers Hsu et al. (1976). Today, there exists a linear homeomorphic muscle model that has the nonlinear properties of muscle, which will be described later.

The development of the material in this chapter is historical, beginning with the earliest models to the more homeomorphic models of today. Before the development of mathematical models, mechanical models of the oculomotor system were created to describe the system. Two illustrations of these models are presented in Fig. 3.1.

3.2 WESTHEIMER SACCADIC EYE MOVEMENT MODEL

The first quantitative saccadic eye movement model was published by Westheimer in 1954. In this model, he simulated horizontal saccades in response to a 20° target displacement. A mechanical

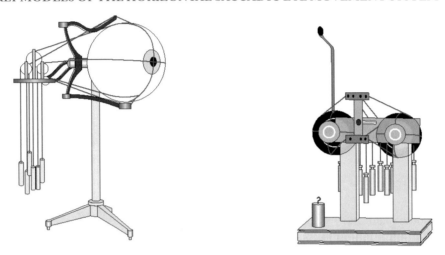

Figure 3.1: The mechanical model by Wundt on the left was used to demonstrate the six muscles of the oculomotor plant, represented as cords, that move the eyeball. The oculomotor model on the right was used to demonstrate the complete eye movement system consisting of the two eyeballs driven by three pairs of muscles each. Illustration adapted from Zimmermann, E. (1903). *XVIII. Preis–Liste über psychologische und psychologische Apparate.*

description of the model is given in Fig. 3.2, and a system description in Eq. (3.1).

$$J\ddot{\theta} + B\dot{\theta} + K\theta = \tau(t) \,. \tag{3.1}$$

To analyze the characteristics of this model and compare it to data, Laplace variable analysis is used. Assuming zero initial conditions, the Laplace transform of Eq. (3.1) yields

$$\theta\left(s^2 J + sB + K\right) = \tau(s) \tag{3.2}$$

and as a transfer function in standard form

$$H(s) = \frac{\theta}{\tau} = \frac{1}{s^2 J + sB + K} = \frac{\frac{\omega_n^2}{K}}{s^2 + 2\zeta\omega_n s + \omega_n^2} \tag{3.3}$$

where according to Westheimer's data for a $20°$ saccade, $\omega_n = \sqrt{\frac{K}{J}} = 120$ and $\zeta = \frac{B}{2\sqrt{KJ}} = 0.7$. According to the data, the roots are complex and given by $s_{1,2} = -\zeta\omega_n \pm j\omega_n\sqrt{1 - \zeta^2} = -84 \pm j85.7$. The input to this system is a step input $\tau(s) = \frac{\gamma}{s}$, where γ is the magnitude of the step input. The response is given by $\theta(s) = H(s)\tau(s)$ or

$$\theta(s) = \frac{\gamma \frac{\omega_n^2}{K}}{s\left(s^2 + 2\zeta\omega_n s + \omega_n^2\right)} \,. \tag{3.4}$$

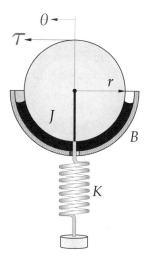

Figure 3.2: A diagram illustrating Westheimer's second-order model of the saccade system. The parameters J, B, and K are rotational elements for moment of inertia, friction, and stiffness, respectively, and represent the eyeball and its associated viscoelasticity. The torque applied to the eyeball by the lateral and medial rectus muscles is given by $\tau(t)$, and θ is the angular eye position. The radius of the eyeball is r.

Equation (3.4) can be expanded via partial fraction expansion to yield

$$\theta(s) = \frac{\frac{\gamma}{K}}{s} + \frac{\frac{\gamma}{2K\left[(\zeta^2-1)-j\zeta\sqrt{1-\zeta^2}\right]}}{\left(s + \zeta\omega_n - j\omega_n\sqrt{1-\zeta^2}\right)} + \frac{\frac{\gamma}{2K\left[(\zeta^2-1)+j\zeta\sqrt{1-\zeta^2}\right]}}{\left(s + \zeta\omega_n + j\omega_n\sqrt{1-\zeta^2}\right)} \tag{3.5}$$

where the magnitude numerator of the complex terms is

$$M = \left| \frac{\gamma}{2K\left[(\zeta^2-1)-j\zeta\sqrt{1-\zeta^2}\right]} \right| = \frac{\gamma}{2K\left\{(\zeta^2-1)^2 + \zeta^2\left(1-\zeta^2\right)\right\}^{\frac{1}{2}}} = \frac{\gamma}{2K\left(1-\zeta^2\right)^{\frac{1}{2}}}$$

and phase angle is

$$\phi = \tan^{-1}\frac{-\zeta\sqrt{1-\zeta^2}}{(\zeta^2-1)} = \tan^{-1}\frac{-\zeta\sqrt{1-\zeta^2}}{-\left(\sqrt{1-\zeta^2}\right)^2} = \tan^{-1}\frac{-\zeta}{-\sqrt{1-\zeta^2}}.$$

The solution to Eq. (3.5) in the time domain is given by

$$\theta(t) = \frac{\gamma}{K}\left[1 + \frac{e^{-\zeta\omega_n t}}{\sqrt{1-\zeta^2}}\cos\left(\omega_n\sqrt{1-\zeta^2}t + \psi\right)\right] \tag{3.6}$$

where $\psi = \pi + \tan^{-1} \dfrac{-\zeta}{\sqrt{1-\zeta^2}}$.

To fully explore the quality of a model, it is necessary to compare its performance against the data. For a saccade, convenient metrics are time to peak overshoot, which gives an indication of saccade duration, and peak velocity. These metrics were discussed previously when the main sequence diagram was described in Section 1.2.

3.2.1   WESTHEIMER'S TIME TO PEAK OVERSHOOT

Next, we wish to solve for time to peak overshoot, T_p, which is important if we want to estimate the parameters of the model. It should be clear that for the Westheimer model, T_p does not change as a function of the input magnitude since the system is linear (i.e., T_p equals a constant for all saccades). The time to peak overshoot is found from

$$\frac{\partial \theta}{\partial t} = \frac{\partial}{\partial t}\left[\frac{\gamma}{K}\left(1 + \frac{e^{-\zeta\omega_n t}}{\sqrt{1-\zeta^2}}\cos\left(\omega_n\sqrt{1-\zeta^2}t + \psi\right)\right)\right]\Bigg|_{t=T_p} = 0 \,. \tag{3.7}$$

Using the chain rule to evaluate Eq. (3.7) and substituting $t = T_p$, yields

$$\frac{\gamma}{K\sqrt{1-\zeta^2}}\left[-\zeta\omega_n\cos(\omega_d T_p + \psi)e^{-\zeta\omega_n t} - \omega_d e^{-\zeta\omega_n t}\sin(\omega_d T_p + \psi)\right] = 0 \,. \tag{3.8}$$

Equation (3.8) is rewritten as

$$-\zeta\cos(\omega_d T_p + \psi) = \sqrt{1-\zeta^2}\sin(\omega_d T_p + \psi)$$

which yields

$$\tan(\omega_d T_p + \psi) = \frac{-\zeta}{\sqrt{1-\zeta^2}} = \tan\psi \,. \tag{3.9}$$

Equation (3.9) is true whenever $\omega_d T_p = n\pi$. The time to peak amplitude is the smallest value that satisfies Eq. (3.9), which is $n = 1$. Thus, $T_p = \frac{\pi}{\omega_d} = \frac{\pi}{\omega_n\sqrt{1-\zeta^2}}$. To determine $\theta(T_p)$, note that $\omega_d t + \psi = \pi + \psi$, and

$$\begin{aligned}
\theta(T_p) &= \frac{\gamma}{K}\left[1 + \frac{e^{\frac{-\zeta\pi}{\sqrt{1-\zeta^2}}}}{\sqrt{1-\zeta^2}}\cos(\pi + \psi)\right] \\
&= \frac{\gamma}{K}\left[1 + \frac{e^{-\frac{\zeta\pi}{\sqrt{1-\zeta^2}}}}{\sqrt{1-\zeta^2}} \times \sqrt{1-\zeta^2}\right] \\
&= \frac{\gamma}{K}\left(1 + e^{\frac{-\pi\zeta}{\sqrt{1-\zeta^2}}}\right) \,.
\end{aligned} \tag{3.10}$$

With Westheimer's constants of $\zeta = 0.7$ and ω_n 120, we find that $T_p = \dfrac{\pi}{120 \times \sqrt{1 - 0.7^2}} = 37$ ms. T_p in the Westheimer model is independent of saccade size and not in agreement with experimental data presented in the main sequence diagram. The data shows a saccade duration that increases with amplitude, where this model has a constant duration.

3.2.2   WESTHEIMERS MAXIMUM VELOCITY

An important aid in examining the suitability of a model is to study the model predictions and data estimates of higher order derivatives. If there are problems with the model, these problems are amplified when comparing estimates of the higher order derivatives with model predictions. For the Westheimer model, maximum velocity is found from

$$\frac{\partial^2 \theta}{\partial t^2}\bigg|_{t=T_{mv}} = 0 \tag{3.11}$$

where T_{mv} is the time at peak velocity. Substituting the solution given by Eq. (3.6) into Eq. (3.11) yields:

$$\begin{aligned}
\frac{\partial^2 \theta}{\partial t^2} &= \frac{\partial}{\partial t}\left[\frac{-\gamma}{K} \frac{e^{-\zeta\omega_n t}}{\sqrt{1-\zeta^2}} \left(\zeta\omega_n \cos(\omega_d t + \psi) + \omega_d \sin(\omega_d t + \psi) \right) \right] \\
&= \frac{-\gamma}{K\sqrt{1-\zeta^2}} \Big[-\zeta\omega_n e^{-\zeta\omega_n t} \left(\zeta\omega_n \cos(\omega_d t + \psi) + \omega_d \sin(\omega_d t + \psi) \right) \\
&\quad + e^{-\zeta\omega_n t} \left(-\zeta\omega_n\omega_d \sin(\omega_d t + \psi) + \omega_d^2 \cos(\omega_d t + \psi) \right) \Big] = 0 \, .
\end{aligned} \tag{3.12}$$

The terms multiplying the sinusoids in Eq. (3.12) are removed since they do not equal zero. Therefore,

$$\left(\omega_d^2 - \zeta^2 \right)\omega_n^2 \cos(\omega_d t + \psi) - 2\zeta\omega_n\omega_d \sin(\omega_d t + \psi) = 0$$

which reduces to

$$\frac{\omega_d^2 - \zeta^2\omega_n^2}{2\zeta\omega_n\omega_d} = \frac{\sin(\omega_d t + \psi)}{\cos(\omega_d t + \psi)} = \tan(\omega_d t + \psi) \, . \tag{3.13}$$

Substituting $\omega_d = \omega_n (1 - \zeta^2)$ into Eq. (3.13), we have

$$\frac{\omega_n^2\left(1 - \zeta^2 \right) - \zeta^2\omega_n^2}{2\zeta\omega_n\omega_n\left(1 - \zeta^2 \right)} = \frac{1 - 2\zeta^2}{2\zeta\sqrt{1 - \zeta^2}} = \tan(\omega_d t + \psi) \, . \tag{3.14}$$

With $\tan\psi = \dfrac{-\zeta}{\sqrt{1-\zeta^2}}$, we factor out $\dfrac{-\zeta}{\sqrt{1-\zeta^2}}$ in Eq. (3.14), substitute $\tan\psi$, giving

$$\frac{1 - 2\zeta^2}{2\zeta\sqrt{1-\zeta^2}} = \frac{-\zeta}{\sqrt{1-\zeta^2}}\left(1 - \frac{1}{2\zeta^2} \right) = \left(1 - \frac{1}{2\zeta^2} \right)\tan\psi = \tan(\omega_d t + \psi) \, . \tag{3.15}$$

Now

$$\tan(\omega_d t + \psi) = \frac{\tan(\omega_d t) + \tan \psi}{1 - \tan(\omega_d t) \tan \psi} \tag{3.16}$$

and

$$\tan \phi = \tan \psi \ .$$

Substituting for $\tan(\omega_d t + \psi)$ from Eq. (3.16) into Eq. (3.15) gives

$$\left(1 - \frac{1}{2\zeta^2}\right) \tan \psi = \frac{\tan(\omega_d t) + \tan \psi}{1 - \tan(\omega_d t) \tan \psi} \ . \tag{3.17}$$

Multiplying both sides of Eq. (3.17) by $(1 - \tan(\omega_d t) \tan \psi)$ gives

$$\left(1 - \frac{1}{2\zeta^2}\right) \tan \psi - \left(1 - \frac{1}{2\zeta^2}\right) \tan(\omega_d t) \tan \psi^2 = \tan(\omega_d t) + \tan \psi \ . \tag{3.18}$$

Collecting like terms in Eq. (3.18) gives

$$\tan(\omega_d t) \left(1 + \tan \psi^2 \left(1 - \frac{1}{2\zeta^2}\right)\right) = \frac{- \tan \psi}{2\zeta^2} \ . \tag{3.19}$$

Dividing both sides of Eq. (3.19) by the term multiplying $\tan(\omega_d t)$ gives

$$\tan(\omega_d t) = \frac{- \tan \psi}{2\zeta^2 \left(1 + \tan \psi^2 \left(\frac{2\zeta^2 - 1}{2\zeta^2}\right)\right)} = \frac{- \tan \psi}{2\zeta^2 + \tan \psi^2 \, 2\zeta^2 - 1} \ . \tag{3.20}$$

With $\tan \psi = \tan \phi = \frac{-\zeta}{\sqrt{1 - \zeta^2}}$, we have

$$\tan(\omega_d t) = \frac{\frac{\zeta}{\sqrt{1 - \zeta^2}}}{2\zeta^2 + \frac{\zeta^2}{1 - \zeta^2} 2\zeta^2 - 1} = \frac{\frac{\zeta}{\sqrt{1 - \zeta^2}}}{\zeta^2 \left(2 + \frac{2\zeta^2 - 1}{1 - \zeta^2}\right)} = \left(\frac{1 - \zeta^2}{1 - \zeta^2}\right) \frac{\frac{\zeta}{\sqrt{1 - \zeta^2}}}{\zeta^2 \left(2 + \frac{2\zeta^2 - 1}{1 - \zeta^2}\right)}$$

$$= \frac{\sqrt{1 - \zeta^2}}{\zeta^2 1 - \zeta^2 + 2\zeta^2 - 1} = \frac{\sqrt{1 - \zeta^2}}{\zeta} \ . \tag{3.21}$$

Taking the inverse tangent of the previous equation gives $T_{mv} = \frac{1}{\omega_d} \tan^{-1} \left(\frac{\sqrt{1 - \zeta^2}}{\zeta}\right)$. $\dot{\theta}(T_{mv})$ can be evaluated using these values for T_{mv}.

Westheimer noted the differences between saccade duration-saccade magnitude and peak velocity-saccade magnitude in the model and the experimental data and inferred that the saccade system was not linear because the peak velocity-saccade magnitude plot was nonlinear. He also noted that the input was not an abrupt step function. Overall, this model provided a satisfactory fit to the eye position data for a saccade of 20°, but it was not for saccades of other magnitudes. Interestingly, Westheimer's second-order model proves to be an adequate model for saccades of all sizes if a different input function, as described in the next section, is assumed. Due to its simplicity, the Westheimer model of the oculomotor plant is still popular today as evidenced by its use in Chapter 2.

3.3   ROBINSON MODEL SACCADIC EYE MOVEMENT MODEL

One of the challenges in modeling physiological systems is the lack of data or information about the input to the system. For instance, in the fast eye movement system the input is the neurological signal from the CNS to the muscles connected to the eyeball. Information about the input is not available in this system since it involves thousands of neurons firing at a very high rate. Recording the signal would involve invasive surgery and instrumentation that was not available in the 1960's. Often times, however, it is possible to obtain information about the input via indirect means as described in this section.

In 1964, Robinson performed an experiment in an attempt to measure the input to the eyeballs during a saccade. To record the input, one eye was held fixed using a suction contact lens, while the other eye performed a saccade from target to target. Since the same innervation signal is sent to both eyes during a saccade, Robinson inferred that the input, recorded through the transducer attached to the fixed eyeball, was the same input driving the other eyeball. He proposed that muscle tension driving the eyeballs during a saccade are a pulse plus a step, or simply, a pulse-step input (Fig. 3.3).

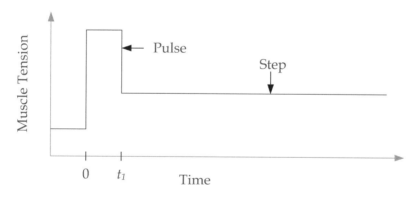

Figure 3.3: Diagram illustrating the muscle tension recorded during a saccade.

Today, microelectrode studies are carried out to record the electrical activity in oculomotor neurons in monkeys. Figure 3.4 illustrates a micropipette being used to record the activity in the oculomotor nucleus, an important neuron population responsible for driving a saccade. Additional experiments on oculomotor muscle have been carried out to learn more about the saccade controller since Robinson's initial study. For instance in 1975, Collins and his co-workers reported using a miniature "C" gauge force transducer to measure muscle tension *in vivo* at the muscle tendon during unrestrained human eye movements. This type of study has allowed a better understanding of the tensions exerted by each muscle, rather than the combined effect of both muscles as shown in Fig. 3.3.

It is important to distinguish between the tension or force generated by a muscle, called muscle tension, and the force generator within the muscle, called the **active-state tension generator**. The

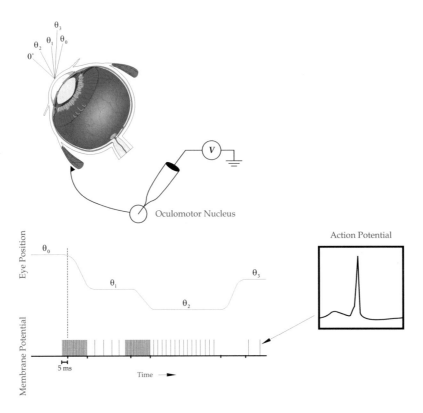

Figure 3.4: Diagram of recording of a series of saccades using a micropipette and the resultant electrical activity in a single neuron. Spikes in the membrane potential indicate an action potential occurred. Saccade neural activity initiates with a burst of neural firing approximately 5 ms before the eye begins to move and continues until the eye has almost reached its destination. Relative position of the eye is shown at the top with angles θ_o through θ_3. Initially, the eye starts in position θ_o, a position in the extremity in which the muscle is completely stretched with zero input. To move the eye from θ_o to θ_1, neural burst firing occurs. To maintain the eye at θ_1, a steady firing occurs in the neuron. The firing rate for fixation is in proportion to the shortness of the muscle. Next, the eye moves from θ_1 to θ_2. This saccade moves much more slowly than the first saccade with approximately the same duration as the first. The firing level is also approximately at the same level as the first. The difference in input corresponds to fewer neurons firing to drive the eye to its destination, which means a smaller input than the first saccade. Because the muscle is shorter after completing this saccade, the fixation firing rate is higher than the previous position at θ_1. Next, the eye moves in the opposite direction to θ_3. Since the muscle is lengthening, the input to the muscle is zero, that is, no action potentials are used to stimulate the muscle. The fixation firing level θ_3 is less than that for θ_1 because the muscle is longer.

active-state tension generator creates a force within the muscle that is transformed through the internal elements of the muscle into the muscle tension. Muscle tension is external and measurable. Active-state tension is internal and not measurable. Active-state tension follows most closely the neural input to the muscle. From Fig. 3.4, a pattern of neural activity is observed as follows:

1. The muscle that is being contracted (agonist) is stimulated by a pulse, followed by a step to maintain the eyeball at its destination.

2. The muscle that is being stretched (antagonist), is unstimulated during the saccade (stimulated by a pause or a negative pulse to zero), followed by a step to maintain the eyeball at its destination.

Figure 3.5 quantifies these relationships for the agonist neural input, N_{ag}, and the antagonist neural input, N_{ant}. The pulse input is required to get the eye to the target as soon as possible, and the step is required to keep the eye at that location. It has been reported that the active-state tensions are not identical to the neural controllers but are described by low-pass filtered pulse-step waveforms. The active-state tensions are shown in Fig. 3.5 as blue lines with time varying time constants τ_{ac} and τ_{de}. It is thought that the low pass filtering involves the movement of Ca^{++} across the cell membrane. Some investigators have reported a different set of time constants for the agonist and antagonist activity, and others have noted a firing frequency dependent agonist activation time constant. Others suggest that the agonist activation time constant is a function of saccade magnitude. For simplicity in this chapter, activation and deactivation time constants are assumed to be identical for both agonist and antagonist activity. The parameters are defined as follows:

F_{go} is the initial agonist active state tension before the saccade starts

F_P is the maximum agonist active state tension

F_{gs} is the steady-state agonist active state tension after the saccade ends

F_{to} is the initial antagonist active state tension before the saccade starts

F_{ts} is the steady-state antagonist active state tension after the saccade ends

Generally, the pulse is used to get the eyeball to the target quickly, and the step is required to keep the eye at that location. The same innervation signal is sent to both eyes, and as a result, they move together. We call this a conjugate eye movement.

In 1964, Robinson described a model for fast eye movements (constructed from empirical considerations) which simulated saccades over a range of 5° to 40° by changing the amplitude of the pulse-step input. These simulation results were adequate for the position–time relationship of a saccade, but the velocity–time relationship was inconsistent with physiological evidence. To correct this deficiency of the model, physiological and modeling studies of the oculomotor plant were carried out during the 1960s through the present that allowed the development of a more **homeomorphic** oculomotor plant. Essential to this work was the construction of oculomotor muscle models. Some of these details are presented in the next section.

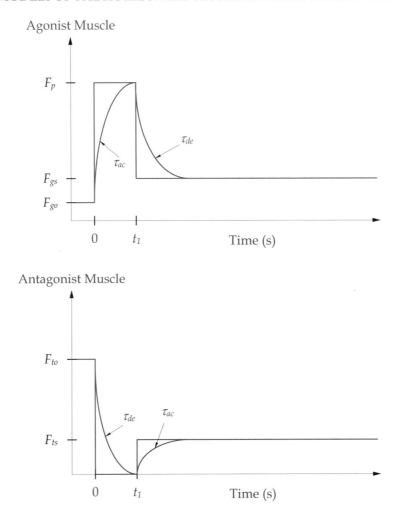

Figure 3.5: Agonist, N_{ag}, and antagonist, N_{ant}, control signals (red lines) and the agonist, F_{ag}, and antagonist, F_{ant}, active-state tensions (blue lines). Note that the time constant for activation, τ_{ac}, is different from the time constant for deactivation, τ_{de}. The time interval, t_1, is the duration of the pulse.

3.4   DEVELOPMENT OF AN OCULOMOTOR MUSCLE MODEL AND THE OCULOMOTOR PLANT

It is clear that an accurate model of muscle is essential for the development of a model of the horizontal fast eye movement system that is driven by a pair of muscles (lateral and medial rectus muscles). In fact, the Westheimer model does not include any muscle model and relies solely on the inertia of the eyeball, friction between the eyeball and socket, and elasticity due to the optic nerve and other attachments as the elements of the model. In this section, the historical development of a muscle model is discussed as it relates to the oculomotor system. Muscle model research involves a broad spectrum of topics, ranging from the micro models that deal with the sarcomeres to macro models, in which collections of cells are grouped into a lumped parameter system and described with ordinary mechanical elements. Here the focus is on a macro model of the oculomotor muscle based on physiological evidence from experimental testing. The model elements, as presented, consist of an active state tension generator (input), elastic elements and viscous elements. Each element is introduced separately and the muscle model is incremented in each subsection. It should be noted that the linear muscle model presented at the end of this section completely revises the subsections before it. The earlier subsections were presented because of their historical significance and to appreciate the current muscle model.

In the last chapter, an accurate oculomotor plant model was not needed since we were focusing on the smooth pursuit system, whose movement is very slow compared to the dynamics of the oculomotor plant. In this chapter, we are focusing on the saccadic system that is directly related to the oculomotor plant dynamics and is of fundamental importance to understanding this system. For this reason, we are interested in creating a very accurate model of muscle that is homeomorphic with the real system as much as possible.

3.4.1   MUSCLE MODEL PASSIVE ELASTICITY

Consider the experiment of stretching an unexcited muscle and recording tension to investigate the passive elastic properties of muscle. The data curve shown in Fig. 3.6 is a typical recording of the tension observed in an eye rectus muscle. The tension required to stretch a muscle is a *nonlinear* function of distance. Thus, in order to precisely model this element, a nonlinear spring element should be used. Note that the change in length at 0 refers to the length of the muscle at primary position (looking straight ahead). Thus, the eye muscles are stretched, approximately 3 mm, when the eye movement system is at rest in primary position. At rest, the muscle length is approximately 37 mm.

To be useful in a linear model of muscle, Fig. 3.6 must be linearized in the vicinity of an operating point. The operating point should be somewhat centered in the region in which the spring operates. In Fig. 3.6, a line tangent to the curve at primary position provides a linear approximation to the spring's behavior in this region as done historically. For ease in analysis, the following relationships hold for a sphere representing the eyeball radius of 11 mm.

$1 \text{ g} = 9.806 \times 10^{-3} \text{ N}$

$1° = 0.192 \text{ mm} = 1.92 \times 10^{-4} \text{ m}$

The slope of the line, K_{pe}, is approximately

$$K_{pe} = 0.2 \, \frac{\text{g}}{°} = 0.2 \, \frac{\text{g}}{°} \times \frac{9.806 \times 10^{-3} \text{ N}}{1 \text{ g}} \times \frac{1°}{1.92 \times 10^{-4}} = 10.2 \, \frac{N}{m}$$

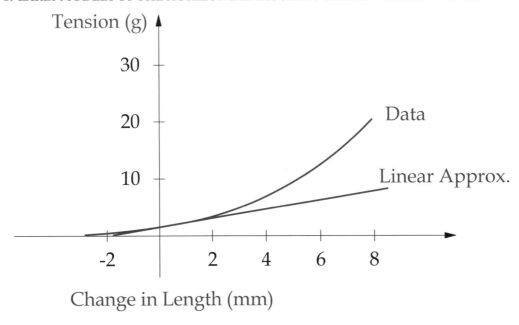

Figure 3.6: Diagram illustrating the tension-displacement curve for unexcited muscle. The slope of the linear approximation to the data is muscle passive elasticity, K_{pe}.

and represents the elasticity of the passive elastic element.

The choice of the operating region is of vital importance concerning the slope of the curve. At this time, a point in the historical operating region of rectus muscle is used. In most of the oculomotor literature, the term K_{pe} is typically subtracted out of the analysis and is not used. The operating point will be revisited in Chapter 5 and this element will be completely removed from the model.

3.4.2   ACTIVE STATE TENSION GENERATOR

In general, a muscle produces a force in proportion to the amount of stimulation. The element responsible for the creation of force is the active state tension generator. Note that this terminology is used so that there is no confusion with the force created within the muscle when the tension created by the muscle is discussed. The active tension generator is included along with the passive elastic element in the muscle model as shown in Fig. 3.7. The relationship between tension, T, active state tension, F, and elasticity is given by

$$T = F - K_{pe}x .$$ (3.22)

Isometric (constant length) experiments have been performed on humans over the years to estimate the active tension generator at different levels of stimulation. These experiments were usually

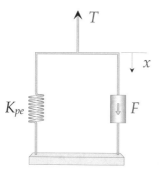

Figure 3.7: Diagram illustrating a muscle model consisting of an active state tension generator, F, and a passive elastic element, K_{pe}. Upon stimulation of the active state tension generator, a tension, T, is exerted by the muscle.

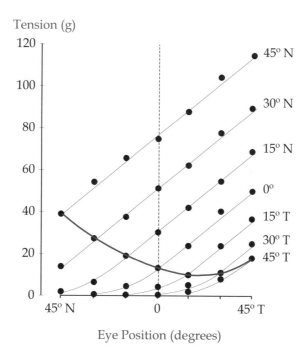

Figure 3.8: Length-tension curves for lateral rectus muscle at different levels of activation. Dots represent tension data recorded from the detached lateral rectus muscle during strabismus surgery while the unoperated eyeball fixated at targets from $-45°$ to $45°$. Adapted from: Collins et al., 1975. Muscle tension during unrestrained human eye movements. *J. Physiol.*, 245: 351.

performed in conjunction with strabismus surgery when muscles were detached and reattached to correct crossed eyes. Consider the tension created by a muscle when stimulated as a function of length as shown in Fig. 3.8. The data were collected from the lateral rectus muscle that was detached from one eyeball while the other unoperated eyeball fixated at different locations in the nasal (N) and temporal (T) directions from −45° to 45°. This experiment was carried out under the assumption that the same neural input is sent to each eyeball (Hering's Law of equal innervation), thus the active state tension in the freely moving eyeball should be the same as that in the detached lateral rectus muscle. At each fixation point, the detached lateral rectus muscle was stretched and tension data was recorded at each of the points indicated on the graph. The thick red line represents the muscle tension at that particular eye position under normal conditions. The curve for 45° T is the zero stimulation case and represents the passive mechanical properties of muscle. Note that the tension generated is a *nonlinear* function of the length of the muscle.

To compare the model in Fig. 3.7 against the data in Fig. 3.8, it is convenient to subtract the passive elasticity in the data (represented by the 45° T curve) from each of the data curves 30° T through 45° N, leaving only the hypothetical active state tension. Shown in the graph on the left in Fig. 3.9 is one such calculation for 15° N with the active state tension given by the dashed line.

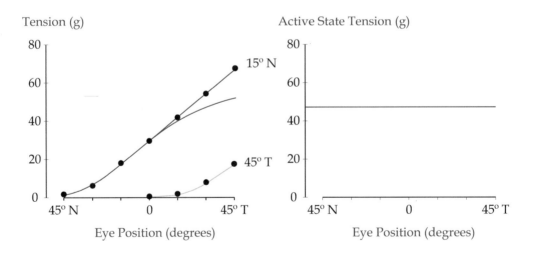

Figure 3.9: On left: length-tension curves for extraocular muscle at two levels of activation corresponding to the 15° N and 45° T positions. The red line represents tension data recorded from the detached lateral rectus muscle during strabismus surgery while the unoperated eyeball fixated at targets. Blue line is the 15° curve with the 45° curve (green line) subtracted from it. The resultant blue curve represents the active state tension as a function of eye position. On right: the theoretical graph for active state tension vs. eye position as given by Eq. (3.22). Adapted from: Collins et al., 1975. Muscle tension during unrestrained human eye movements. *J. Physiol.*, 245: 351.

The other curves in Fig. 3.8 give similar results and have been omitted because of the clutter. The blue line should represent the active state tension, which appears to be a function of length. If this was a *pure* active state tension element, the subtracted curve should be a horizontal line indicative of the size of the input. One such input is shown for the active state tension in the graph on the right in Fig. 3.9. The result in Fig. 3.9 implies that either the active state tension's effect is a nonlinear element, i.e., there may be other nonlinear or linear elements missing in the model, or, perhaps, some of the assumptions made in the development of the model are wrong. For the moment, consider that the analysis is correct and assume that some elements are missing. This topic will be revisited at the end of this chapter.

3.4.3   ELASTICITY

The normal operating point (at primary position), L_p, is much shorter than the length at which maximum force occurs at approximately 30°. Even when the effects of the passive muscle are removed, a relationship between length and tension is still evident, as previously described. Let us introduce a new elastic element into the model to account for the relationship between length and tension as shown in Fig. 3.10, which is described by the following equation.

$$T = F - K_{pe}x - Kx .$$

(3.23)

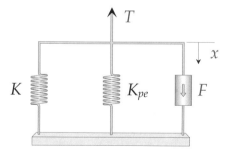

Figure 3.10: Diagram illustrating a muscle model consisting of an active state tension generator F, passive elastic element, K_{pe}, and elastic element, K. Upon stimulation of the active state tension generator, a tension T is exerted by the muscle.

The new elastic element, K, accounts for the slope of the subtracted curve shown with in blue in the graph on the left in Fig. 3.9. The slope of the line, K, at primary position, is approximately 0.8 g/° = 40.86 N/m (a value typically reported in the literature). The slope for each of the curves in Fig. 3.8 can be calculated in the same manner at primary position with the resultant slopes all approximately equal to the same value as the one for 15° N. At this time, the introduction of additional experiments will provide further insight to the development of the muscle model.

SERIES ELASTIC ELEMENT

Experiments carried out by Levin and Wyman in 1927, and Collins in 1975 indicated the need for a series elasticity element, in addition to the other elements previously presented in the muscle model. The experimental setup and typical data from the experiment are shown in Fig. 3.11. The protocol for this experiment, called the quick release experiment, is as follows. (1) A weight is hung onto the

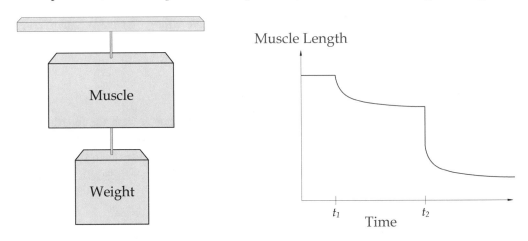

Figure 3.11: Diagram illustrating the quick release experiment. Figure on left depicts the physical setup of the experiment. Figure on the right shows typical data from the experiment. At time t_1 the muscle is fully stimulated and at time t_2 the weight is released.

muscle. (2) The muscle is fully stimulated at time t_1. (3) The weight is released at time t_2. At time t_2, the muscle changes length almost instantaneously when the weight is released. The only element that can instantaneously change its length is a spring. Thus, to account for this behavior, a spring, called the series elastic element, K_{se}, is connected in series to the active state tension element. While some investigators argue that this element is nonlinear, we will assume that it is linear for simplicity. The updated muscle model is shown in Fig. 3.12.

    Based on the experiment carried out by Collins, C. (1975) on rectus eye muscle, an estimate for K_{se} was given as 125 N/m (2.5 gm/°). Since the value of K_{se} does not equal the value of K, another elastic element is needed to account for this behavior.

LENGTH TENSION ELASTIC ELEMENT

Given the inequality between K_{se} and K, another elastic element, called the length-tension elastic element, K_{lt}, is placed in parallel with the active state tension element as shown in the illustration on the left in Fig. 3.13. For ease of analysis, K_{pe} is subtracted out (removed) using the graphical technique shown in Fig. 3.9. To estimate a value for K_{lt}, the muscle model shown on the right in Fig. 3.13 is analyzed and reduced to an expression involving K_{lt}. Analysis begins by summing the

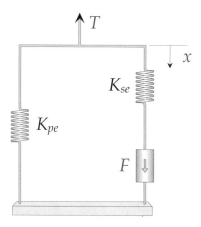

Figure 3.12: Diagram illustrating a muscle model consisting of an active state tension generator F, passive elastic element K_{pe}, and series elastic element K_{se}. Upon stimulation of the active state tension generator F, a tension T is exerted by the muscle.

forces acting on nodes 1 and 2.

$$T = K_{se}(x_2 - x_1) \tag{3.24}$$

$$F = K_{lt}x_2 + K_{se}(x_2 - x_1) \qquad \rightarrow \qquad x_2 = \frac{F + K_{se}x_1}{K_{se} + K_{lt}} . \tag{3.25}$$

Substituting x_2 from Eq. (3.25) into (3.24) gives

$$T = \frac{K_{se}}{K_{se} + K_{lt}}(F + K_{se}x_1) - K_{se}x_1 = \frac{K_{se}}{K_{se} + K_{lt}}F - \frac{K_{se}K_{lt}}{K_{se} + K_{lt}}x_1 . \tag{3.26}$$

Equation (3.26) is an equation for a straight line with y-intercept $\frac{K_{se}}{K_{se}+K_{lt}}F$ and slope $\frac{K_{se}K_{lt}}{K_{se}+K_{lt}}$. The slope of the length-tension curve in Fig. 3.9 is given by $K = 0.8$ g/$^\circ$ = 40.86 N/m. Therefore,

$$K = \frac{K_{se}K_{lt}}{K_{se} + K_{lt}} = 40.86 \quad \frac{N}{m} . \tag{3.27}$$

Solving Eq. (3.27) for K_{lt} yields

$$K_{lt} = \frac{K_{se}K}{K_{se} - K} = 60.7 \quad \frac{N}{m} . \tag{3.28}$$

3.4.4 FORCE-VELOCITY RELATIONSHIP

Early experiments indicated that muscle had elastic as well as viscous properties. Muscle was tested under isotonic (constant force) experimental conditions as shown in Fig. 3.14 to investigate muscle

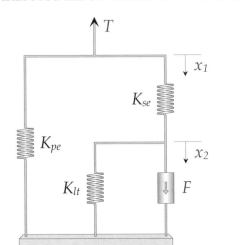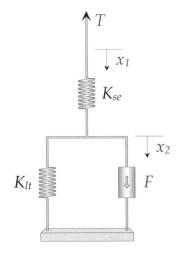

Figure 3.13: Diagram on left illustrates a muscle model consisting of an active state tension generator F in parallel to a length-tension elastic element K_{lt}, connected to a series elastic element K_{se}, all in parallel with the passive elastic element K_{pe}. Upon stimulation of the active state tension generator F, a tension T is exerted by the muscle. The diagram on the right is the same muscle model except that K_{pe} has been removed.

viscosity. The muscle and load were attached to a lever with a high lever ratio. The lever reduced the gravity force (mass × gravity) of the load at the muscle by one over the lever ratio, and the inertial force (mass × acceleration) of the load by one over the lever ratio squared. With this arrangement, it was assumed that the inertial force exerted by the load during isotonic shortening could be ignored. The second assumption was that if mass was not reduced enough by the lever ratio (enough to be ignored), then taking measurements at maximum velocity provided a measurement at a time when acceleration is zero, and, therefore, inertial force equals zero. If these two assumptions are valid, then the experiment would provide data free of the effect of inertial force as the gravity force is varied.

According to the experimental conditions, the muscle is stretched to its optimal length at the start of the isotonic experiment. The isotonic experiment begins by attaching a load M, stimulating the muscle, and recording position. The two curves in Fig. 3.15 depict the time course for the isotonic experiment for a small and large load. Notice that the duration of both responses are approximately equal regardless of the load, in spite of the apparent much longer time delay associated with the large load. Next, notice that the heavier the load, the less the total shortening. Maximum velocity is calculated numerically from the position data. To estimate muscle viscosity, this experiment is repeated with many loads at the same stimulation level, and maximum velocity is calculated. Figure 3.16 illustrates the typical relationship between load ratio (P/P_o) and maximum velocity,

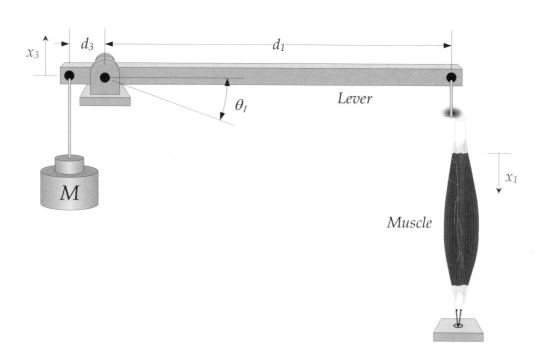

Figure 3.14: Drawing of the classical isotonic experiment with inertial load and muscle attached to the lever. The muscle is stretched to its optimal length according to experimental conditions and attached to ground.

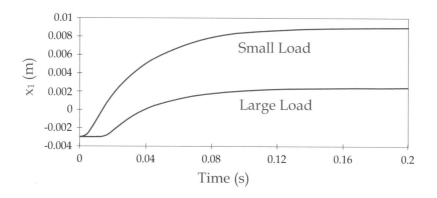

Figure 3.15: Diagram illustrating typical response of a muscle stimulated with a large and small load.

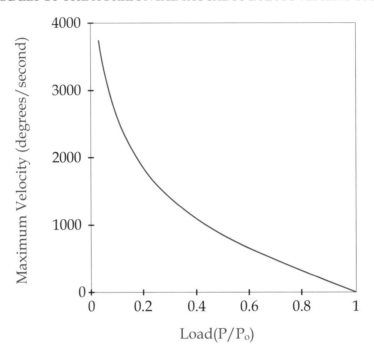

Figure 3.16: Illustrative force-velocity curve.

where $P = Mg$ and P_o is the isometric tension (the largest weight that the muscle can move) for maximally stimulated muscle. This curve is usually referred to as the **force-velocity curve**.

Clearly, the force-velocity curve is nonlinear and follows a hyperbolic shape. If a smaller stimulus than maximum is used to stimulate the muscle, then a family of force-velocity curves results as shown in Fig. 3.17. Each curve is generated with a different active state tension as indicated. The force-velocity characteristics in Fig. 3.17 are similar to those shown in Fig. 3.16. In particular, the slope of the force-velocity curve for a small value of active state tension is quite different than that for a large value of the active state tension in the operating region of the eye muscle (i.e., approximately 800 °/s).

To include the effects of viscosity from the isotonic experiment in the muscle model, a viscous element is placed in parallel with the active state tension generator and the length tension elastic element as shown in Fig. 3.18. The impact of this element is examined by analyzing the behavior of the model in Example 3.1 by simulating the conditions of the isotonic experiment. At this stage, it is assumed that the viscous element is linear in this example. For simplicity, the lever is removed along with the virtual acceleration term $M\ddot{x}_1$. A more thorough analysis including the lever is considered later in this book. For simplicity, the passive elastic element K_{pe} is removed from the diagram and analysis.

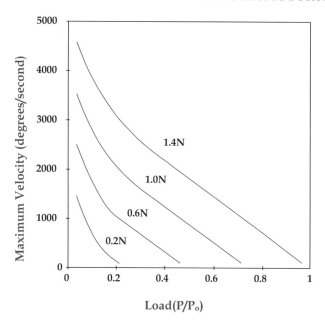

Figure 3.17: Illustrative family of force-velocity curves for active state tensions ranging from 1.4 to 0.2 N.

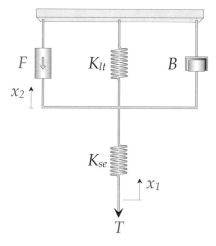

Figure 3.18: Diagram illustrates a muscle model consisting of an active state tension generator F, in parallel with a length-tension elastic element K_{lt}, and viscous element B, connected to a series elastic element K_{se}. The passive elastic element K_{pe} has been removed from the model for simplicity. Upon stimulation of the active state tension generator F, a tension T is exerted by the muscle.

Example 3.1
Consider the system shown in Fig. 3.19 that represents a model of the isotonic experiment. Assume that the virtual acceleration term $M\ddot{x}_1$ can be ignored. Calculate and plot maximum velocity as a function of load.

Solution.
Assume that $\dot{x}_2 > \dot{x}_1$, and that the mass is supported so that $x_1 > 0$. Let the term $K_{st} = K_{se} + K_{lt}$. Summing the forces acting on nodes 1 and 2 gives

$$Mg = K_{se}(x_2 - x_1) \qquad \rightarrow \qquad x_1 = x_2 - \frac{Mg}{K_{se}}$$
$$F = B\dot{x}_2 + K_{lt}x_2 + K_{se}(x_2 - x_1).$$

Substituting x_1 into the second equation yields

$$F = B\dot{x}_2 + K_{lt}x_2 + Mg.$$

Solving the previous equation for x_2 and $\dot{x}_2$ gives

$$x_2(t) = \frac{F - Mg}{K_{lt}}\left(1 - e^{-\frac{K_{lt}t}{B}}\right)$$

$$\dot{x}_2(t) = \frac{F - Mg}{B}e^{-\frac{K_{lt}t}{B}}.$$

Maximum velocity, V_{max}, for all loads is given by $V_{max} = \frac{F-Mg}{B}$ and $\dot{x}_1 = \dot{x}_2$ since $\dot{x}_1 = \frac{d}{dt}\left(x_2 - \frac{Mg}{K_{se}}\right)$. Figure 3.20 depicts a linear relationship between maximum velocity and load.

The assumption of a linear viscosity element appears to be in error since the analysis in Example 3.1 predicts a linear relationship between load and maximum velocity (according to the assumptions of the solution), and the data from the isotonic experiment shown in Fig. 3.16 is clearly nonlinear. Thus, a reasonable assumption is that the viscosity element is nonlinear.

Traditionally, muscle viscosity is characterized by the nonlinear Hill hyperbola, given by

$$V_{max}(P + a) = b(P_0 - P) \tag{3.29}$$

where V_{max} is the maximum velocity, P is the external force, P_o the isometric tension, and a and b are the empirical constants representing the asymptotes of the hyperbola. As described previously, P_o represents the isometric tension, which is the largest weight that the muscle can move, and P is the weight Mg. Hill's data suggests that

$$a = \frac{P_0}{4} \qquad \text{and} \qquad b = \frac{V_{max}}{4}.$$

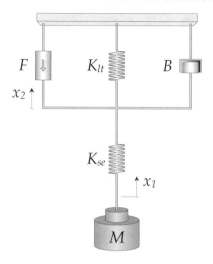

Figure 3.19: System for Example 3.1.

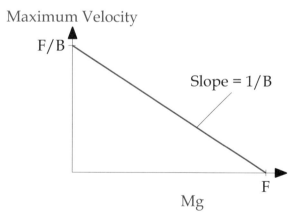

Figure 3.20: Result for Example 3.1.

Therefore, with these values for a and b, the Hill equation is rewritten from Eq. (3.29) as

$$P = P_0 - \frac{V_{\max}(P_0 + a)}{b + V_{\max}} = P_0 - B V_{\max} \tag{3.30}$$

where

$$B = \frac{P_0 + a}{b + V_{\max}}. \tag{3.31}$$

The term B represents the viscosity of the element. Clearly, the force due to viscosity is nonlinear due to the velocity term, V_{max}, in the denominator of Eq. (3.30).

In oculomotor models, V_{max} is usually replaced by $\dot{x}_2$, P is replaced by muscle tension, T, and P_o is replaced by the active state tension, F, as defined from Fig. 3.19. Therefore, Eqs. (3.30) and (3.31) are rewritten as

$$T = F - BV \tag{3.32}$$

where

$$B = \frac{F + a}{b + \dot{x}_2} \ . \tag{3.33}$$

Some oculomotor investigators have reported values for a and b in the Hill equation that depend on whether the muscle is being stretched or contracted. There is some evidence to suggest that stretch dynamics are different from contraction dynamics. However, the form of the viscosity expression for muscle shortening or lengthening is given by Eq. (3.33), with values for a and b parameterized appropriately. For instance, Hsu and coworkers described the viscosity for shortening and lengthening for oculomotor muscles as

$$B_{ag} = \frac{F_{ag} + AG_a}{\dot{x}_2 + AG_b} \tag{3.34}$$

$$B_{ant} = \frac{F_{ant} - ANT_a}{-\dot{x}_2 - ANT_b} \tag{3.35}$$

where AG_a, AG_b, ANT_a, and ANT_b are parameters based on the asymptotes for contracting (agonist) or stretching (antagonist), respectively.

3.4.5   MUSCLE MODEL

At this time, we will put all of the elements that have been discussed into a model of muscle as shown in Fig. 3.21 (Left), and then we will analyze this model to determine the tension created by the muscle. Note that in the muscle model, we have subtracted out the effects of passive elasticity and assumed that $\dot{x}_2 > \dot{x}_1$. Thus, starting with the free body diagram in Fig. 3.21 (Right), we have our two node equations as

$$T = K_{se} (x_2 - x_1)$$
$$F = B\dot{x}_2 + K_{lt}x_2 + K_{se} (x_2 - x_1) \ .$$

We solve for x_2 from the node 1 equation as $x_2 = \frac{T}{K_{se}} + x_1$, and we substitute it into the node 2 equation, giving us

$$F = B\dot{x}_2 + (K_{se} + K_{lt}) x_2 - K_{se}x_1 = B\dot{x}_2 + K_{st} \left(\frac{T}{K_{se}} + x_1 \right) - K_{se}x_1 \ .$$

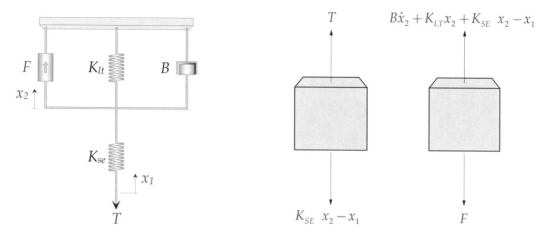

Figure 3.21: (Left) Updated model of muscle with active state tension generator, length-tension elastic element, series elastic element and viscosity element. (Right) Free body diagram of the system on the left.

For convenience, we will let $K_{st} = K_{se} + K_{lt}$, and we will multiply the previous equation by K_{se} and rearrange terms, thus, we have

$$K_{se}F = K_{se}B\dot{x}_2 + K_{st}T + K_{se}K_{lt}x_1$$

or

$$T = \frac{K_{se}F}{K_{st}} - \frac{K_{se}K_{lt}}{K_{st}}x_1 - \frac{K_{se}B}{K_{st}}\dot{x}_2 \, .$$

Equation (3.33), or those in Eqs. (3.34) and (3.35), depending on whether the muscle is contracting or lengthening, can be substituted for parameter B in the above equation to give a nonlinear model of oculomotor muscle.

3.4.6   PASSIVE TISSUES OF THE EYEBALL

At this point, we return to modeling the eyeball. As previously discussed, Robinson not only described the passive properties of muscle, he also determined the elasticity, viscosity, and inertia of the eyeball from his experiments during strabismus surgery. With the two horizontal muscles, Fig. 3.22 describes the passive tissues of the eyeball.

Note that the passive elasticity of the eyeball, K, is a combination of the effects due to the four other muscles, optic nerve, etc. The viscous element of the eyeball, B, is due to the friction of the eyeball within the eye socket.

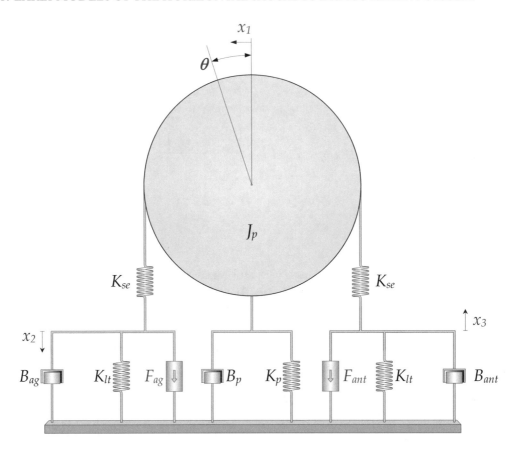

Figure 3.22: This model describes the two rectus muscles (agonist (ag) and antagonist (ant)), connected to the eyeball through nodes 1 and 4. θ represents the angle that the eyeball is rotated and x_1 represents the length of arc rotated. Variables x_2 and x_3 represent the length of the muscles.

3.4.7   ACTIVATION AND DEACTIVATION TIME CONSTANTS

The control signal that the central nervous system sends to the oculomotor system during a saccade is a pulse-step signal as described in Fig. 3.5. The signal the oculomotor system actually experiences is a low pass filtered version of this signal as represented by the blue lines in Fig. 3.5.

If we let $C(s)$ = control signal, $F(s)$ = active state tension and $H(s)$ = low-pass filter, then

$$F(s) = C(s) H(s) = \frac{C(s)}{(s\tau + 1)}$$

where τ is the low-pass filter time constant.

The agonist time constant reported by Bahill, A. (1980) is a function of motoneuronal firing frequency, the higher the rate, the shorter the time constant. As explained, this is because large saccadic eye movements utilize fast muscle fibers and smaller saccadic eye movements utilize slow muscle fibers. There are two muscles involved with a horizontal eye movement, the agonist and antagonist muscles. The agonist muscle forcibly contracts and moves the eyeball (fovea) to the target location. The antagonist muscle is completely inhibited during the pulse phase of the trajectory. Keep in mind that muscles are always under stimulation (tonic state at primary position) to avoid slack.

The control signal from the CNS to each muscle is a series of pulses or spikes due to the action potentials of the motoneurons, as illustrated in Fig. 3.4. This diagram illustrates a typical pattern observed during a series of fast eye movements in both horizontal directions. Notice that during a movement in the "on" direction (lateral), the rate of firing increases greatly; in the "off" direction (medial), the firing rate is zero. Also, notice that the burst firing starts approximately 5 ms before the saccade begins, and that the longer the neurons fire, the larger the saccade.

There is a large, nonconstant time delay between the time a target moves, and when the eye actually starts to move. This is due to the *CNS* system calculating the forces necessary to bring the fovea to the target location. This movement is *ballistic* (not guided) to the extent that there are no known stretch receptors indicating muscle activity.

3.5   1976 NONLINEAR RECIPROCAL INNERVATION SACCADE MODEL

Significant strides have been made in modeling the oculomotor plant with nonlinear and linear models and the neural network controlling the eye movements. Despite this progress in oculomotor research, much work remains in the development of an oculomotor muscle model that can be used in systems applications and that exhibits realistic characteristics of rectus eye muscle.

At this time, we will derive a nonlinear model of the fast eye movement system based on the work of Hsu et al. (1976) using the model in Fig. 3.22. The nonlinearity of the model is due to the viscosity elements of the muscle as discussed in Section 3.4.4.

To begin our analysis of the saccades for horizontal eye movements only, we assume that:

1. $\dot{x}_2 > \dot{x}_1 > \dot{x}_3$

2. elasticity K_p is the passive elasticity from the superior and inferior oblique muscles, the superior and inferior rectus muscles and the eyeball

3. x_i is measured in mm from the equilibrium position

4. zero initial conditions

Note that $x_1 = \theta r$ or $\theta = \frac{x_1}{r} = 5.2087 \times 10^3 x_1$, where x_1 is measured in meters. To begin the analysis, we draw free body diagrams as shown in Fig. 3.23, and we write the node equations and

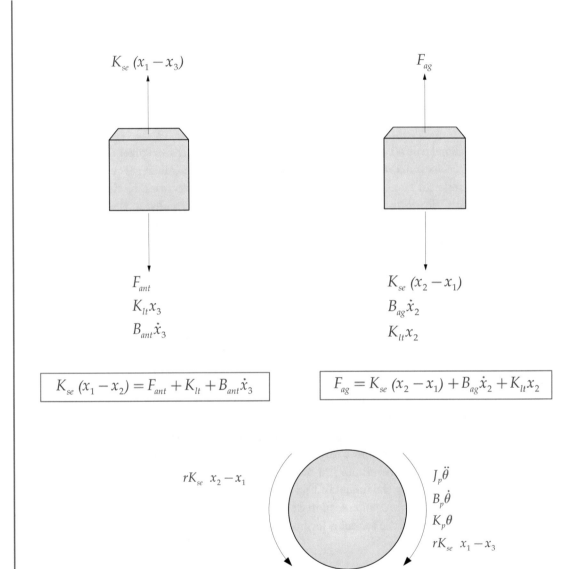

Figure 3.23: Free body diagrams for nonlinear reciprocal innervation model.

sum of torques about the eyeball as

$$K_{se}(x_1 - x_2) = F_{ant} + K_{lt} + B_{ant}\dot{x}_3$$
$$F_{ag} = K_{se}(x_2 - x_1) + B_{ag}\dot{x}_2 + K_{lt}x_2 \qquad (3.36)$$
$$r K_{se}(x_2 - x_1) = J_p\ddot{\theta} + B_p\dot{\theta} + K_p\theta + r K_{se}(x_1 - x_3) \ .$$

We substitute $\theta = \frac{x_1}{r}$ into Eq. (3.36), giving

$$r K_{se}(x_2 + x_3 - 2x_1) = \frac{J_p}{r}\ddot{x}_1 + \frac{B_p}{r}\dot{x}_1 + \frac{K_p}{r}x_1$$
$$K_{se}(x_1 - x_3) = F_{ant} + K_{lt}x_3 + B_{ant}\dot{x}_3$$
$$F_{ag} = K_{se}(x_2 - x_1) + B_{ag}\dot{x}_2 + K_{lt}x_2 \ .$$

Consistent with the original model development by Bahill, we assume $K_{lt} = 0$ and let $J = \frac{J_p}{r^2}$, $B = \frac{B_p}{r^2}$, and $K = \frac{K_p}{r^2}$. After making these substitutions, we have

$$K_{se}(x_2 + x_3 - 2x_1) = J\ddot{x}_1 + B\dot{x}_1 + Kx_1$$
$$K_{se}(x_1 - x_3) = F_{ant} + B_{ant}\dot{x}_3 \qquad (3.37)$$
$$F_{ag} = K_{se}(x_2 - x_1) + B_{ag}\dot{x}_2 \ .$$

Based on the work of Hsu et al. (1976), we write the following nonlinear viscosity terms for our model

$$B_{ag} = \frac{F_{ag} + AG_a}{\dot{x}_2 + AG_b} \qquad B_{ant} = \frac{F_{ant} - ANT_a}{-\dot{x}_3 - ANT_b} \ .$$

Substituting these terms into our model gives us

$$K_{se}(x_2 + x_3 - 2x_1) = J\ddot{x}_1 + B\dot{x}_1 + Kx_1$$
$$K_{se}(x_1 - x_3) = F_{ant} + B_{ant}\dot{x}_3 = F_{ant} + \left(\frac{F_{ant} - ANT_a}{-\dot{x}_3 - ANT_b}\right)\dot{x}_3 \qquad (3.38)$$
$$F_{ag} = K_{se}(x_2 - x_1) + B_{ag}\dot{x}_2 = K_{se}(x_2 - x_1) + \left(\frac{F_{ag} + AG_a}{\dot{x}_2 + AG_b}\right)\dot{x}_2 \ .$$

Since B_{ag} and B_{ant} are nonlinear, we cannot directly solve the differential equations or use Laplace transforms to solve for θ. The results for θ must be simulated. Here, we use SIMULINK, and for convenience, we redefine the model into a state variable format. Let the state variables be x_1, v_1, x_2, x_3, F_{ag}, and F_{ant}.

For state variable x_1, by inspection, we write $\dot{x}_1 = v_1$.

For state variable x_2, we use the node equation for the agonist muscle and solve for $\dot{x}_2$, giving

$$F_{ag} = K_{se}\,(x_2 - x_1) + B_{ag}\dot{x}_2 = K_{se}\,(x_2 - x_1) + \left(\frac{F_{ag} + AG_a}{\dot{x}_2 + AG_b}\right)\dot{x}_2$$

$$(\dot{x}_2 + AG_b)\,F_{ag} = K_{se}\,(x_2 - x_1)\,(\dot{x}_2 + AG_b) + \left(F_{ag} + AG_a\right)\dot{x}_2$$

$$\dot{x}_2 = \frac{AG_b\left(F_{ag} - K_{se}\,(x_2 - x_1)\right)}{AG_a + K_{se}\,(x_2 - x_1)}\,.$$

For state variable x_3, we use the node equation for the antagonist muscle and solve for $\dot{x}_3$, giving

$$K_{se}\,(x_1 - x_3) = F_{ant} + B_{ant}\dot{x}_3 = F_{ant} + \left(\frac{F_{ant} - ANT_a}{-\dot{x}_3 - ANT_b}\right)\dot{x}_3$$

$$(-\dot{x}_3 - ANT_b)\,K_{se}\,(x_1 - x_3) = (-\dot{x}_3 - ANT_b)\,F_{ant}\dot{x}_3 - ANT_a\dot{x}_3$$

$$\dot{x}_3 = \frac{ANT_b\,(K_{se}\,(x_1 - x_3) - F_{ant})}{ANT_a - K_{se}\,(x_1 - x_3)}\,.$$

For state variable v_1, we use the node equation for the eyeball, substitute $\dot{v}_1 = \ddot{x}_1$ and $v_1 = \dot{x}_1$, and solve for $\dot{v}_1$, giving

$$\dot{v}_1 = \frac{1}{J}\left[K_{se}\,(x_2 + x_3 - 2x_1) - Bv_1 - Kx_1\right]\,.$$

For the last two state variables, we define them as low-pass filtered neural inputs as

$$\dot{F}_{ag} = \frac{N_{ag} - F_{ag}}{\tau_{ag}}$$

$$\dot{F}_{ant} = \frac{N_{ant} - F_{ant}}{\tau_{ant}}\,.$$

3.5.1 PARAMETERS FOR THE NONLINEAR RECIPROCAL INNERVATION MODEL

The following are parameter values for the nonlinear reciprocal innervation model that simulate saccades of all sizes.

$K_{se} = 1.8$ g tension/degree = 91.9 N/m

$K = 0.86$ g tension/degree = 43.9 N/m

$B = 0.018$ g tension-sec/degree = 0.919 N-s/m

$J = 4.3 \times 10^{-5}$ g tension-s^2/degree = 2.192 $\times 10^{-3}$ N-s/m

$N_{\text{AG-Pulse}} = $ PH (see Table 3.1)

$N_{ANT-Pulse} = \left(0.5 + 16e^{-\frac{\Delta\theta}{2.5}}\right)$ g tension

$N_{AG-Step} = (16 + 0.8\Delta\theta)$ g tension

$N_{ANT-Step} = (16 - 0.06\Delta\theta)$ g tension

$\tau_{AG-AC} = (13 - 0.1\Delta\theta)$ ms

$\tau_{AG-DE} = 2$ ms

$\tau_{ANT-AC} = 3$ ms

$\tau_{ANT-DE} = 11$ ms

Agonist Pulse Width (PW) (see Table 3.1)

Antagonist PW circumscribes Agonist PW by 3 ms on each side

Table 3.1: Agonist Pulse width and height for 7 saccades ranging from .1 to 30 degrees.

Magnitude (degree)	Pulse Width, PW, (ms)	Pulse Height, PH, (g)
0.1	10	17.6
0.5	10	20
1	11	22
5	15	53
10	20	87
20	31	124
30	40	155

For the previous parameters, the nonlinear force-velocity relationships are given as:

$$B_{ag} = \frac{F_{ag} + AG_a}{\dot{x}_2 + AG_b} = \frac{1.25 F_{ag}}{\dot{x}_2 + 900} \quad \frac{g-s}{\circ}$$

$$B_{ant} = \frac{F_{ant} - ANT_a}{-\dot{x}_3 - ANT_b} = \frac{F_{ant} - 40}{-\dot{x}_3 - 900} \quad \frac{g-s}{\circ}.$$

Note that:

1. 1 g tension $= 9.806 \times 10^{-3}$ N

2. $1° = 1.92 \times 10^{-4}$ m

$\Delta\theta$ represents the absolute value of the eye movement size.

Example 3.2

Simulate a 10° saccade using the 1976 nonlinear reciprocal innervation oculomotor model. Plot the neural inputs, agonist and antagonist active-state tension, position, velocity and acceleration vs. time.

Solution.
Using the values in Table 3.1, the following m-file automatically provides the agonist pulse width and height. The parameter `zero` is the start time for the saccade, here set at 10 ms.

```
%dTheta: Input parameter
dTheta=input('dTheta Value in degrees:')
zero=10e-3;
switch dTheta
    case 0.1, PH=17.6; PW=10e-3;
    case 0.5, PH=20;   PW=10e-3;
    case 1,   PH=22;   PW=11e-3;
    case 5,   PH=53;   PW=15e-3;
    case 10,  PH=87;   PW=20e-3;
    case 20,  PH=124;  PW=31e-3;
    case 30,  PH=155;  PW=40e-3;
end;
N_AG_Pulse=PH;
N_AG_Step=16+0.8*dTheta;
N_AN_Pulse=0.5+16*exp(-dTheta/2.5);
N_AN_Step=16-0.06*dTheta;
TauAG_AC=(13-0.1*dTheta)*1e-3;
```

Shown in Fig. 3.24 is the Simulink program. The main program is shown in Fig. 3.24 (A). The agonist and antagonist input are shown in Fig. 3.24 (B) and (C). Equation (3.38) is implemented in Fig. 3.24 (D). In Fig. 3.25 are plots of position, velocity, acceleration, agonist neural input and active state tension, and antagonist neural input and active state tension.

3.6   1984 LINEAR RECIPROCAL INNERVATION OCULOMOTOR MODEL

Based on physiological evidence, Bahill et al. (1980) presented a linear 4th order model of the horizontal oculomotor plant that provides an excellent match between model predictions and horizontal eye movement data. This model eliminates the differences seen between velocity and acceleration predictions of the Westheimer and Robinson models and the data. For ease in this presentation, the 1984 modification of this model by Enderle and coworkers is used.

In the previous analysis, B_{ag} and B_{ant} are nonlinear functions of velocity. We can linearize these functions by approximating the force-velocity family of curves with straight line segments as illustrated in Fig. 3.26. Antagonist activity is typically at the 5% level and agonist activity is at the

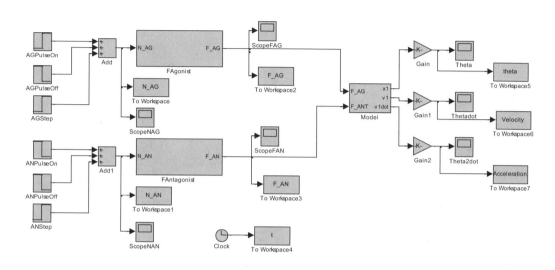

Figure 3.24: Simulink program for Example 3.2 (A) (Main Simulink Program).

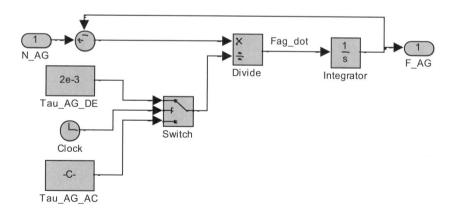

Figure 3.24: (continued). Simulink program for Example 3.2 (B) (Agonist Input).

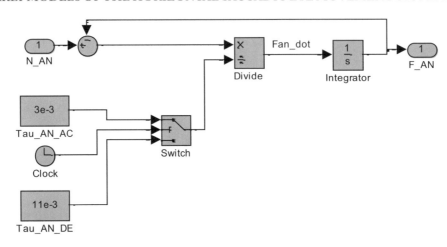

Figure 3.24: (continued). Simulink program for Example 3.2 (C) (Antagonist Input).

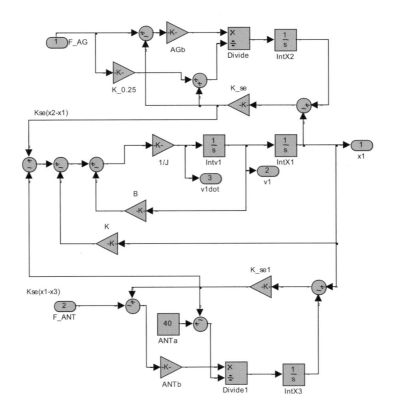

Figure 3.24: (continued). Implementation of Eq. (3.38) (D).

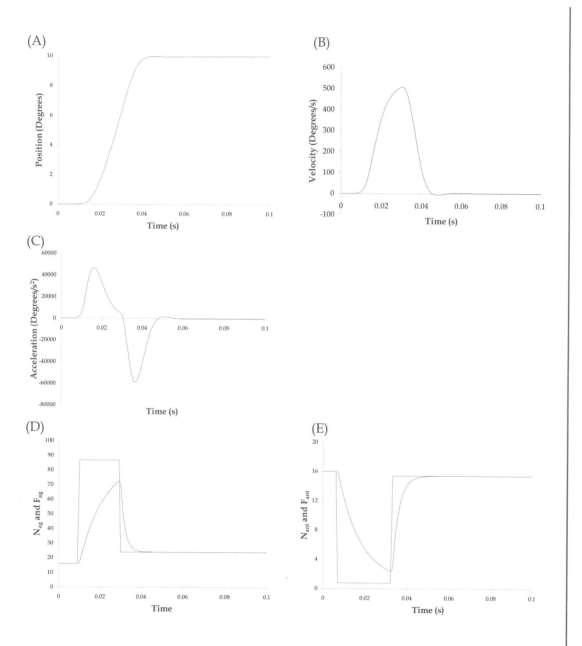

Figure 3.25: Plots of position, velocity, acceleration, agonist neural input and active state tension, and antagonist neural input and active state tension for Example 3.2.

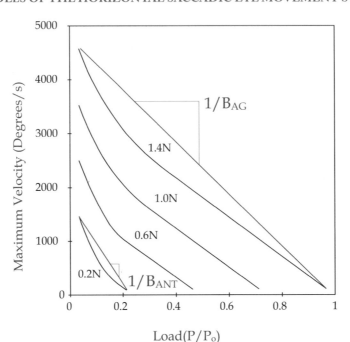

Figure 3.26: Linearization of nonlinear force-velocity curves shown in red.

100% level. Thus, we can assume that B_{ag} and B_{ant} are constants with different values since the slopes are different in the linearization.

Using the linearized viscosity elements in our model of the eye movement system, we will now derive a linear differential equation describing saccades as a function of θ. The updated model is shown in Fig. 3.27. The material presented here is based on the work published by Bahill and his coworkers (Bahill et al., 1980), and Enderle and coworkers (Enderle et al., 1984).

To begin the analysis, we first draw the free body diagrams and write the node equations as shown in Fig. 3.28.

Node 1: $rK_{se}(x_2 - x_1) - rK_{se}(x_4 - x_3) = J_p\ddot{\theta} + B_p\dot{\theta} + K_p\theta$

Node 2: $F_{ag} = B_{ag}\dot{x}_2 + K_{se}(x_2 - x_1) + K_{lt}x_2$

Node 3: $K_{se}(x_4 - x_3) = F_{ant} + K_{lt}x_3 + B_{ant}\dot{x}_3$

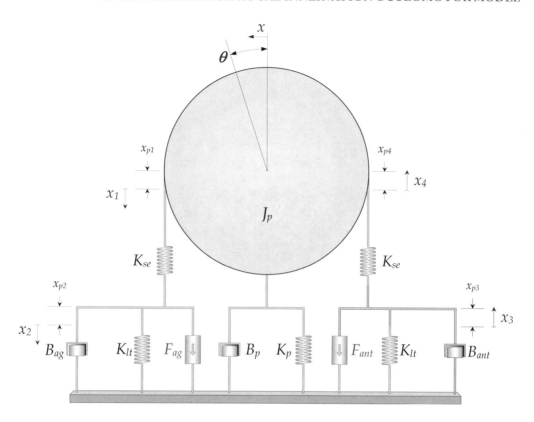

Figure 3.27: Linear eye movement model.

Next, we let

$$J = \frac{J_p}{r} \times 5.2087 \times 10^3, \; B = \frac{B_p}{r} \times 5.2087 \times 10^3$$

$$K = \frac{K_p}{r} \times 5.2087 \times 10^3, \; \text{and } \theta = \frac{x}{r} \times \frac{180}{\pi} = 5.2087 \times 10^3 x$$

and rewrite the node 1 equation as

$$K_{se} (x_2 + x_3 - x_1 - x_4) = J\ddot{x} + B\dot{x} + Kx \; .$$

We assume that there is an initial displacement from equilibrium at primary position for springs K_{lt} and K_{se}, since the muscle is 3 mm longer than at equilibrium. That is

$$x_1 = x - x_{p1}$$
$$x_4 = x + x_{p4} \; .$$

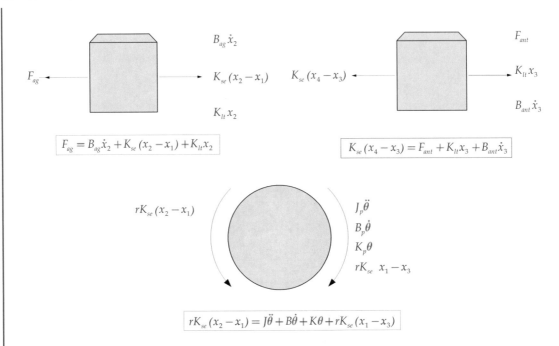

Figure 3.28: Free body diagrams for the system in Fig. 3.27.

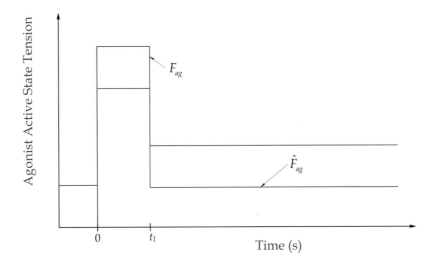

Figure 3.29: Illustration of the relationship between F_{ag} (red line) and $\hat{F}_{ag}$ (blue line).

To reduce the node equations to a single differential equation, we must eliminate variables. We use the operating point analysis method, and we introduce the following variables:

$$\hat{x} = x - x\,(0)$$
$$\hat{\theta} = \theta - \theta\,(0)$$
$$\hat{x}_1 = x_1 - x_1\,(0)$$
$$\hat{x}_2 = x_2 - x_2\,(0)$$
$$\hat{x}_3 = x_3 - x_3\,(0)$$
$$\hat{x}_4 = x_4 - x_4\,(0)$$
$$\hat{F}_{ag} = F_{ag} - F_{ag}\,(0)$$
$$\hat{F}_{ant} = F_{ant} - F_{ant}\,(0)$$
$$K_{st} = K_{se} + K_{lt}$$

Note that $\hat{x} = \hat{x}_1 = \hat{x}_4$. To appreciate the relationship among the variables, note that if F_{ag} is a pulse-step, then $\hat{F}_{ag}$ is shown in Fig. 3.29.

We now need to determine the relationship among variables at time zero, recognizing that derivative terms are zero. Our three-node equations at steady-state are written as:

$$F_{ag}\,(0) = K_{st}x_2\,(0) - K_{se}x_1\,(0)$$

$$F_{ant}\,(0) = -K_{st}x_3\,(0) + K_{se}x_4\,(0) \qquad (3.39)$$

$$K_{se}\,(x_2\,(0) + x_3\,(0) - x_1\,(0) - x_4\,(0)) = Kx(0) = 0\,.$$

Subtracting the muscle node equations for use later, gives:

$$F_{ag}\,(0) - F_{ant}\,(0) = K_{st}\,(x_2\,(0) + x_3\,(0)) - K_{se}\,(x_1\,(0) + x_4\,(0))\,. \qquad (3.40)$$

We now substitute the operating point variables and initial conditions, yielding

$$\hat{F}_{ag} + F_{ag}\,(0) = K_{st}\left(\hat{x}_2 + x_2\,(0)\right) + B_{ag}\dot{\hat{x}}_2 - K_{se}\left(\hat{x}_1 + x_1\,(0)\right)\,.$$

Removing the initial condition terms with our steady-state analysis, gives

$$\hat{F}_{ag} = K_{st}\hat{x}_2 + B_{ag}\dot{\hat{x}}_2 - K_{se}\hat{x}$$

where $\hat{x}_1$ has been replaced by $\hat{x}$. After repeating this for the other two equations, we have

$$\hat{F}_{ant} = K_{se}\hat{x} - K_{st}\hat{x}_3 - B_{ant}\dot{\hat{x}}_3$$

$$K_{se}\left(\hat{x}_2 + \hat{x}_3 - 2\hat{x}\right) = J\ddot{\hat{x}} + B\dot{\hat{x}} + K\hat{x}$$

where $\hat{x}_4$ has been replaced by $\hat{x}$. We next apply the Laplace transform on the previous three equations, giving

$$\hat{F}_{ag}(s) = \hat{X}_2 \left(K_{st} + s B_{ag} \right) - K_{se}\hat{X} \tag{3.41}$$

$$\hat{F}_{ant}(s) = K_{se}\hat{X} - \hat{X}_3 \left(K_{st} + s B_{ant} \right) \tag{3.42}$$

$$K_{se} \left(\hat{X}_2 + \hat{X}_3 - 2\hat{X} \right) = \left(J s^2 + B s + K \right) \hat{X} \tag{3.43}$$

Rearranging Eqs. (3.41) and (3.42) yields

$$\hat{X}_2 = \frac{\hat{F}_{ag}(s) + K_{se}\hat{X}}{K_{st} + s B_{ag}}$$

$$\hat{X}_3 = \frac{K_{se}\hat{X} - \hat{F}_{ant}(s)}{K_{st} + s B_{ant}}$$

and after substituting $\hat{X}_2$ and $\hat{X}_3$ into Eq. (3.43), we have

$$K_{se} \left(\frac{\left(\hat{F}_{ag}(s) + K_{se}\hat{X} \right)}{\left(K_{st} + s B_{ag} \right)} + \frac{\left(K_{se}\hat{X} - \hat{F}_{ant}(s) \right)}{\left(K_{st} + s B_{ant} \right)} - 2\hat{X} \right) = \left(s^2 J + B s + K \right) \hat{X} .$$

Next, we multiply the previous equation by $(s B_{ant} + K_{st}) (s B_{ag} + K_{st})$, giving us

$$K_{se} \left[(s B_{ant} + K_{st}) \, \hat{F}_{ag}(s) - (s B_{ag} + K_{st}) \, \hat{F}_{ant}(s) \right] = \left(P_4 s^4 + P_3 s^3 + P_2 s^2 + P_1 s + P_0 \right) \hat{X}$$

where

$P_4 = J B_{ant} B_{ag}$
$P_3 = J K_{st} \left(B_{ag} + B_{ant} \right) + B B_{ant} B_{ag}$
$P_2 = J K_{st}^2 + B K_{st} \left(B_{ag} + B_{ant} \right) + B_{ag} B_{ant} \left(K + 2 K_{se} \right)$
$P_1 = B K_{st}^2 + \left(B_{ag} + B_{ant} \right) \left(K K_{st} + 2 K_{se} K_{st} - K_{se}^2 \right)$
$P_0 = K K_{st}^2 + 2 K_{se} K_{st} K_{lt}.$

We now transform back into time domain using the inverse Laplace transform, yielding

$$K_{se} \left(K_{st} \left(\hat{F}_{ag} - \hat{F}_{ant} \right) + B_{ant} \dot{\hat{F}}_{ag} - B_{ag} \dot{\hat{F}}_{ant} \right) = P_4 \, \ddddot{\hat{x}} + P_3 \, \dddot{\hat{x}} + P_2 \, \ddot{\hat{x}} + P_1 \, \dot{\hat{x}} + P_0 \hat{x} . \tag{3.44}$$

Since $\dot{F}_{ag} = \dot{\hat{F}}_{ag}$ and $\dot{F}_{ant} = \dot{\hat{F}}_{ant}$, we have

$$\hat{F}_{ag} - \hat{F}_{ant} = F_{ag} - F_{ag}(0) - F_{ant} + F_{ant}(0) . \tag{3.45}$$

From Eq. (3.40), we have

$$F_{ag}(0) - F_{ant}(0) = K_{st}(x_2(0) + x_3(0)) - K_{se}(x_1(0) + x_4(0))$$

and

$$x_1(0) = x(0) - x_{p1} \text{ and } x_4(0) = x(0) + x_{p1} .$$

Assuming identical muscles, gives us

$$F_{ag}(0) - F_{ant}(0) = K_{st}(x_2(0) + x_3(0)) - 2K_{se}x(0) . \tag{3.46}$$

We have from Eq. (3.39),

$$K_{se}(x_2(0) + x_3(0) - 2x(0)) = Kx(0)$$

or

$$x_2(0) + x_3(0) = \left(\frac{K}{K_{se}} + 2\right)x(0)$$

and when substituted into Eq. (3.46), gives

$$F_{ag}(0) - F_{ant}(0) = \left(K_{st}\left(\frac{K}{K_{se}} + 2\right) - 2K_{se}\right)x(0) . \tag{3.47}$$

With Eqs. (3.45) and (3.46) inserted into Eq. (3.44), we have

$$K_{se}\left[K_{st}\left(F_{ag} - F_{ant}\right) - K_{st}\left(F_{ag}(0) - F_{ant}(0)\right) + B_{ant}\dot{F}_{ag} - B_{ag}\dot{F}_{ant}\right]$$
$$= P_4\,\dddot{x} + P_3\dddot{x} + P_2\ddot{x} + P_1\dot{x} + P_0(x - x(0)) . \tag{3.48}$$

To reduce Eq. (3.48) further, we note, using Eq. (3.47), that

$$\begin{aligned}
K_{se}K_{st}\left(F_{ag}(0) - F_{ant}(0)\right) &= K_{se}K_{st}\left(K_{st}\left(\frac{K}{K_{se}} + 2\right) - 2K_{se}\right)x(0) \\
&= \left(K_{st}^2 K + 2K_{se}K_{st}^2 - 2K_{se}^2 K_{st}\right)x(0) \\
&= \left(K_{st}^2 K + 2K_{se}K_{st}K_{lt}\right)x(0) \\
&= P_0 x(0)
\end{aligned}$$

and substituting this result back into Eq. (3.48) gives

$$K_{se}\left\{K_{st}\left(F_{ag} - F_{ant}\right) + B_{ant}\dot{F}_{ag} - B_{ag}\dot{F}_{ant}\right\} = P_4\,\dddot{x} + P_3\dddot{x} + P_2\ddot{x} + P_1\dot{x} + P_0 x .$$

Exchanging x with θ yields:

$$\delta\left(K_{st}\left(F_{ag} - F_{ant}\right) + B_{ant}\dot{F}_{ag} - B_{ag}\dot{F}_{ant}\right) = \ddddot{\theta} + C_3\dddot{\theta} + C_2\ddot{\theta} + C_1\dot{\theta} + C_0\theta \tag{3.49}$$

where

$$\delta = \frac{57.296 K_{se}}{r J B_{ant} B_{ag}}$$

$$C_3 = \frac{J K_{st} \left(B_{ag} + B_{ant}\right) + B B_{ant} B_{ag}}{J B_{ant} B_{ag}}$$

$$C_2 = \frac{J K_{st}^2 + B K_{st} \left(B_{ag} + B_{ant}\right) + B_{ag} B_{ant} \left(K + 2 K_{se}\right)}{J B_{ant} B_{ag}}$$

$$C_1 = \frac{B K_{st}^2 + \left(B_{ag} + B_{ant}\right) \left(K K_{st} + 2 K_{se} K_{st} - K_{se}^2\right)}{J B_{ant} B_{ag}}$$

$$C_0 = \frac{K K_{st}^2 + 2 K_{se} K_{st} K_{lt}}{J B_{ant} B_{ag}}.$$

A block diagram for the oculomotor plant is shown in Fig. 3.30. The section of the diagram

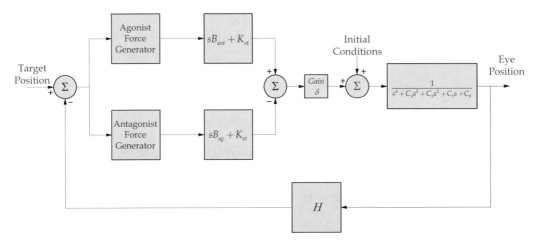

Figure 3.30: Block diagram of the modified linear homeomorphic eye movement model.

in the forward path is the linear homeomorphic model. The feedback element H is unity and is operational only when the eye is not executing a saccade.

3.6.1 METHODS

Data are usually collected from subjects seated before a target display of small light emitting diodes (LEDs), each separated by five degrees (see Fig. 1.3). The subject's head is restrained by a bit-bar, and the subject is instructed to follow the "jumping" target as it moves from the center position to any of the other LEDs and then returns to the center position. The order of the target positions as well as the time interval between displacements are randomized. Data are recorded only for the initial displacement from the center position.

Horizontal eye movements are recorded from each eye using an infrared signal reflected from the anterior surface of the cornea-scleral interface with standard instrumentation (see Fig. 1.3). Signals for both eyes tracking are digitized using the analog/digital converter and stored in disk memory. These signals are typically sampled at a rate of 1000 samples per second for one-half second after the target has moved.

3.6.2   SYSTEM IDENTIFICATION AND VALIDATION

A theoretical model's validity must be established by comparison with experimental data if confidence in its predictive capabilities is expected. Parameters for the theoretical model are determined from physiological evidence or numerical estimation. Simulations are performed with a reliable parameter set and compared with the experimental data. A close match between the data and simulation results under a variety of conditions lend support to the theoretical model's validity. In 1988, Enderle and Wolfe published a paper (Enderle and Wolfe, 1988) in which they used the system identification technique to estimate the oculomotor plant parameters and the agonist and antagonist active state tension during a series of saccades from three subjects using the model presented in this section.

The agonist and antagonist active state tensions are given by the following low-pass filtered pulse-step waveforms, and they are illustrated along with their corresponding neurological control signals in Fig. 3.5.

$$F_{ag} = F_{g0}u(-t) + \left(F_p + \left(F_{g0} - F_p \right) e^{\frac{-t}{\tau_{ac}}} \right) (u(t) - u(t - t_1))$$
$$+ \left(F_{gs} + \left(F_p + \left(F_{g0} - F_p \right) \right) e^{\frac{-t_1}{\tau_{ac}}} - F_{gs} \right) e^{\frac{-(t-t_1)}{\tau_{de}}} u(t - t_1)$$
$$F_{ant} = F_{t0}u(-t) + F_{t0}e^{\frac{-t}{\tau_{de}}} (u(t) - u(t - t_1))$$
$$+ \left(F_{ts} + \left(F_{t0}e^{\frac{-t_1}{\tau_{de}}} - F_{ts} \right) e^{\frac{-(t-t_1)}{\tau_{ac}}} \right) u(t - t_1) \tag{3.50}$$

where

F_{g0} = initial magnitude of the agonist active state tension
F_p = pulse magnitude of the agonist active state tension
F_{gs} = step magnitude of the agonist active state tension
F_{t0} = initial magnitude of the antagonist active state tension
F_{ts} = step magnitude of the antagonist active state tension
τ_{ac} = activation time constant
τ_{de} = deactivation time constant
t_1 = duration of the agonist pulse active state tension

Final parameter estimates for the saccadic eye movement model are found using the system identification technique, a frequency response method. The oculomotor system operates in an open-loop mode while executing a saccade. After completing the saccade, the central nervous system operates in a closed loop mode and compares eye and target position (Carpenter, R., 1988). Figure 3.30 presents a block diagram illustrating the open-loop, closed-loop operation of the oculomotor system, with the feedback element H operating only during discrete-time intervals when a saccade is not being executed. Our system identification is used to estimate parameters and the input during the open loop mode of the saccade.

For the oculomotor system, the transfer function is calculated from the fast eye response to a step in target displacement. Unequal time delays for eye displacements from saccade to saccade for the same target displacement and variability in time to peak velocity and peak velocity make it impossible to use averaging techniques to reduce the effects of measurement noise. Fortunately, the measurement noise is small relative to the input and output signals, and, therefore, the transfer function is calculated as the ratio of the Fourier transform of a single saccadic eye movement and the input signal as follows. First, the fast eye response measurements are filtered using a 4th order Butterworth digital low-pass filter with a half-power point at 125 Hz. Transforming the filtered measurements directly by the fast Fourier algorithm resulted in distortion due to truncation since the signal did not go to zero at steady state. This is circumvented by subtracting the steady-state value from each sample and passing this signal through a Kaiser window. Next, this modified data sequence is extended to a total length of 2048 samples to increase the frequency domain resolution and to force the data length to a power of two (necessary for the fast Fourier transform). Note that subtracting the steady-state value from each sample makes it more convenient to extend the sequence. This modified data sequence is then transformed using the fast Fourier algorithm. Note that the Fourier transform of the fast eye response is now equal to the fast Fourier transform of the modified signal plus the Fourier transform of the unit step with amplitude equal to the steady-state value. The input signal is the Fourier transform of the unit step function with amplitude equal to the steady-state value. Parameter estimates for the oculomotor model in Eq. (3.49) and the active state tensions in Eq. (3.50) are calculated using the conjugate gradient search program similar to Seidel's (Seidel, R., 1975), which minimizes the integral of the absolute value of the error squared between the model and the data.

Initial estimates of the oculomotor mechanical components are based on published data, and initial estimates of the oculomotor control signals are based on an extrapolation of the eye movement trajectory. Specifically, the oculomotor saccade model is solved for the time to peak velocity and peak velocity to yield detailed control signal information. The system identification technique is used to obtain final estimates for the oculomotor mechanical components and control signals by minimizing the quadratic distance between the modified linear homeomorphic model and the data. Although time domain techniques have been used to estimate oculomotor mechanical components and control signals by other investigators, the advantage of this approach is that it allows an estimate of the transfer function, actually the Fourier transform of the output-input ratio, and permits a validity test in both the time and frequency domain.

Great care is exercised in evaluating initial parameter estimates and active state tensions since large differences from the true values could cause the estimation routine to converge to suboptimal and nonphysiologically consistent results. The initial estimates for the mechanical components C_i and δ in Eq. (3.49) are specified from published experimental data. Details on oculomotor muscle elasticity and viscosity, and static active state tensions are presented in this section. As described, a new method of muscle viscosity linearization is presented. The initial estimates for the dynamic active state tensions are specified through theoretical peak velocity results and manual data extrapolation.

MUSCLE VISCOSITY

The nonlinear force-velocity relationship established by isotonic experiments by Fenn and Marsh (1935), quantitatively analyzed by Hill, A. (1938), and linearized by Bahill et al. (1980), is re-examined here. Fig. 3.31 illustrates a linear model of muscle, at equilibrium with a weight Mg

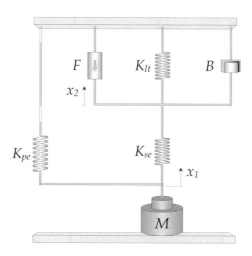

Figure 3.31: Muscle connected to a weight.

attached, which depicts the isotonic experiment. The equations describing the forces acting at junctions 1 and 2 are

$$Mg + K_{pe}x_1 = K_{se}(x_2 - x_1)$$
$$F = B\dot{x}_2 + (K_{se} + K_{lt})x_2 - K_{se}x_1 . \tag{3.51}$$

Note that we have not included the $M\ddot{x}_1$ term to be consistent with the original experiment. By eliminating x_2 from Eq. (3.51), we have

$$F - \frac{(K_{se} + K_{lt})}{K_{se}}Mg = B\left(\frac{K_{pe}}{K_{se}} + 1\right)\dot{x}_1 + \left(K_{lt} + \frac{(K_{se} + K_{lt})K_{pe}}{K_{se}}\right). \tag{3.52}$$

Solving Eq. (3.52) gives

$$x_1 = \frac{\left(F - \frac{(K_{se}+K_{lt})Mg}{K_{se}}\right)}{\left(K_{lt} + \frac{(K_{se}+K_{lt})K_{pe}}{K_{se}}\right)}\left(1 - e^{\frac{-\left(K_{lt} + \frac{(K_{se}+K_{lt})K_{pe}}{K_{se}}\right)t}{B\left(\frac{K_{pe}}{K_{se}}+1\right)}}\right)$$

$$\dot{x}_1 = \frac{\left(F - \frac{(K_{se}+K_{lt})Mg}{K_{se}}\right)}{B\left(\frac{K_{pe}}{K_{se}}+1\right)}e^{\frac{-\left(K_{lt} + \frac{(K_{se}+K_{lt})K_{pe}}{K_{se}}\right)t}{B\left(\frac{K_{pe}}{K_{se}}+1\right)}}. \tag{3.53}$$

Since maximum velocity occurs at time zero, we have

$$V_{\max} = \dot{x}_1(0) = \frac{\left(F - \frac{(K_{se}+K_{lt})Mg}{K_{se}}\right)}{B\left(\frac{K_{pe}}{K_{se}}+1\right)} \tag{3.54}$$

which is a function of Mg according to the isotonic experimental procedure since F is a constant equal to 100% stimulation. Figure 3.32 displays the weight vs. maximum velocity relationship for

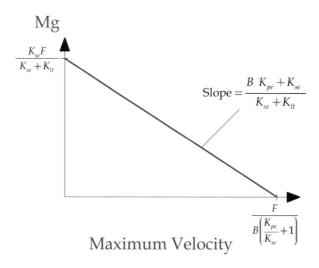

Figure 3.32: Weight vs. maximum velocity relationship for the linear muscle model.

the linear muscle model with slope

$$m = \frac{B\left(K_{pe} + K_{se}\right)}{(K_{se} + K_{lt})}. \tag{3.55}$$

Since the experimental data is hyperbolic, these results are linearized in the vicinity of the operating point and the viscosity B determined by Eq. (3.55). Using the equations for agonist and antagonist viscosity provided by Hsu et al. (1976), the linearized agonist and antagonist viscosities are determined by substituting the following empirical relationships for m in Eq. (3.55):

$$m_{ag} = \frac{1.25 F_{ag}}{AG_b + V_{max}} \quad gs/°$$

$$m_{ant} = \frac{F_{ant} - 40}{-V_{max} - AT_b} \quad gs/° \tag{3.56}$$

where $F_{ag} = 87$ g, $F_{ant} = 0.79$ g, and AG_b and AT_b are constants associated with Hill's hyperbolic equation. Since the values for AG_b and AT_b were updated by Robinson, D. (1981), we use $AG_b = AT_b = 1422°s^{-1}$ instead of the values from Hsu et al. (1976).

Viscosity estimates are computed by linearizing in the vicinity of the maximum saccade velocity. For instance, a $10°$ saccade with $\dot{\theta}(t_{mv}) = 432°s^{-1}$ (from Bahill et al., 1981) has

$$m_{ag} = 3.0 \text{ Nsm}^{-1}$$
$$m_{ant} = 1.1 \text{ Nsm}^{-1}$$
$$B_{ag} = 3.4 \text{ Nsm}^{-1}$$
$$B_{ag} = 1.2 \text{ Nsm}^{-1}$$

using Eqs. (3.55) and (3.56) with $K_{pe} = 12.26 \text{ Nm}^{-1}$ (from Robinson et al., 1969). Table 3.2 gives a list of values of the agonist and antagonist muscle viscosities as a function of saccade magnitude (computed with $\dot{\theta}(t_{mv})$ from Bahill et al., 1981). These values are based on a maximum attainable peak velocity of $684°s^{-1}$ and would be correspondingly smaller as this value is increased i.e., some investigators indicate maximum attainable peak velocities in the range of $800 - 1000°s^{-1}$).

Other investigators have not included the relationship of Eq. (3.55) for the estimates of agonist and antagonist viscosity or linearizing on the basis of $\dot{\theta}(t_{mv})$. Bahill et al. (1980) report $B_{ag} = 3.0 \text{ Nsm}^{-1}$ and $B_{ant} = 1.4 \text{ Nsm}^{-1}$ (after adjusting their incorrectly derived results by the factor $(K_{se} + K_{lt})/K_{se}$) (Enderle et al., 1984). Further adjusting of Bahill estimates using Eq. (3.55) raises the viscosities to $B_{ag} = 3.4 \text{ Nsm}^{-1}$ and $B_{ant} = 1.6 \text{ Nsm}^{-1}$, which closely corresponds to the values in Table 3.2. Other researchers report significantly different linearized muscle viscosity estimates. Lehman and Stark (1979) report $B_{ag} = 0.9 \text{ Nsm}^{-1}$ and $B_{ant} = 2.4 \text{ Nsm}^{-1}$ (after using Eq. (3.55)), based on average value of the nonlinear viscosity Lehman and Stark (1979). In a parametric sensitivity analysis of the oculomotor system, Hsu et al. (1976) and Lehman and Stark (1982) identified the muscle viscosity as having a relatively strong effect on saccadic response. Enderle et al. further established the importance of muscle viscosity on the saccadic response while examining the linear homeomorphic saccadic eye movement model (Enderle et al., 1984). They report that the muscle viscosities multiply the rate of change of the muscle forces in the saccadic eye movement differential equation, thus the muscle viscosities contribute to the forces that drive the eyeball to its destination (Enderle et al., 1984).

Table 3.2: Agonist and antagonist viscosities as a function of saccade magnitude.

Magnitude Degree	$\dot{\theta}(t_{mv})$ °s^{-1}	B_{ag} Nsm^{-1}	B_{ant} Nsm^{-1}
5	269	3.8	1.4
10	432	3.4	1.2
15	531	3.3	1.2
20	591	3.2	1.1

In examining the main sequence diagram, two factors are particularly significant in regard to muscle viscosity. First, since $\dot{\theta}(t_{mv})$ is a function of the saccade magnitude, agonist and antagonist viscosity are also functions of saccade magnitude. Second, since maximum velocity variability exists for saccadic eye movements of the same magnitude, variability also exists for estimates of agonist and antagonist viscosity, which, in turn, causes variability in the estimates for the eigenvalues of the oculomotor system.

OCULOMOTOR MUSCLE ELASTICITY

At steady state, the tensions applied to the eyeball are

$$T_{ag} - T_{ant} = K_p x \tag{3.57}$$

where

T_{ag} = agonist muscle tension minus passive elasticity.
T_{ant} = antagonist muscle tension minus passive elasticity.
From Robinson et al. (1969, Fig. 23), the following relationships are extrapolated with length corrected slopes

$$T_{ag} = 17 \text{ g} + \frac{1 \text{ g}}{\text{degree}} \theta \tag{3.58}$$

and

$$T_{ant} = 17 \text{ g} - \frac{0.3 \text{ g}}{\text{degree}} \theta . \tag{3.59}$$

Substituting Eqs. (3.58) and (3.59) into Eq. (3.57) yields $K_p = 66.4 \text{ Nm}^{-1}$. Note, one can either use tension data to calculate K_p, or use physiologic data on the passive elasticity of the two muscles plus the eyeball. That is, passive elasticity that acts past primary position is 40.35 Nm^{-1} [antagonist muscle], plus passive elasticity that acts short of primary position is 6.64 Nm^{-1} [agonist muscle], plus passive elasticity from the eyeball is 24.52 Nm^{-1}, which yields 71.51 Nm^{-1} (Robinson, D., 1981). We use the value $K_p = 66.4 \text{ Nm}^{-1}$ in our simulations. Lehman and Stark (1982) identified the passive elasticity as a critical parameter in fitting a model to saccadic eye movement data. Other researchers have estimated $K_p = 25$ Nm^{-1} (Bahill et al., 1980), 44 Nm^{-1} (Hsu et al., 1976), 76.6 Nm^{-1} (Cook and Stark, 1967), and 97 Nm^{-1} (Cook and Stark, 1968).

The length tension elasticity is estimated from the slope of the length tension curve from Robinson et al. (1969, Fig. 23). The slope of the length-tension curve, K', is the series combination of the series elasticity and length tension elasticity elements, with the muscle passive elasticity removed.

$$K' = \frac{K_{se}K_{lt}}{K_{lt} + K_{se}} = 25.5 \text{ Nm}^{-1} \tag{3.60}$$

at primary position from the 0° curve. From quick release experiments, Collins estimates $K_{se} = 125$ Nm^{-1} (Collins, C., 1975). Therefore, from Eq. (3.60)

$$K_{lt} = \frac{K_{se}K'}{K_{se} - K'} = 32 \text{ Nm}^{-1} . \tag{3.61}$$

STATIC ACTIVE STATE TENSIONS

In his paper, Collins, C. (1975) reported only the value for the series elastic element, K_{se}, from the isotonic-isometric quick release experiment without explanation. May has described the experiment and verified the results using the muscle model of the modified linear homeomorphic model (May, A., 1985). The experiment was performed on the medial rectus muscle detached at the globe end of a patient undergoing strabismus surgery. With the unoperated eye, the patient looked 15° nasally, which resulted in a corresponding set of active state tensions in the operated eye. The medial rectus muscle was initially stretched to 9 mm beyond slack and quick released a total of 1 mm. In the operating region of the experiment, the passive elasticity of the muscle is negligible and is eliminated from the analysis. Figure 3.33 illustrates a model of the muscle depicting the quick release experiment, where T is muscle tension. Using D'Alembert's law yields

$$T(t) = K_{se}(x_2(t) - x_1(t)) . \tag{3.62}$$

Diagrams (b) and (c) in Fig. 3.33 contain the experimental results from Collins' experiment. Since x_1 is the total change in the muscle length and x_2 cannot change instantaneously, we have

$$T(0^-) - T(0^+) = K_{se}\left(x_1\left(0^+\right) - x_1\left(0^-\right)\right)$$

or
$$K_{se} = \frac{T(0^-) - T(0^+)}{K_{se}\left(x_1\left(0^+\right) - x_1\left(0^-\right)\right)} = 125 \text{ Nm}^{-1} . \tag{3.63}$$

The estimates of the static active state tensions during fixation must satisfy Eq. (3.49), that is

$$\delta(K_{se} + K_{lt})\left(F_{ag} - F_{ant}\right) = C_0\theta . \tag{3.64}$$

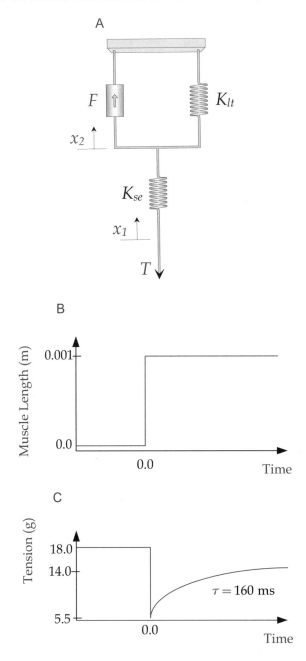

Figure 3.33: (a) Model of muscle depicting the quick release experiment where T is the tension exerted by the muscle; (b)-(c) diagrams illustrating the experimental results from Collins et al. (1975) experiment.

To eliminate the active state tensions during fixation, the following equations describe the forces acting at nodes 1-4 in Fig. 3.26.

$$T_{ag} = K_{se} (x_2 - x_1) \tag{3.65}$$
$$F_{ag} = K_{lt}x_2 + T_{ag} \tag{3.66}$$
$$T_{ant} = K_{se} (x_4 - x_3) \tag{3.67}$$
$$F_{ant} = T_{ant} - K_{lt}x_3 . \tag{3.68}$$

The static agonist active state tension is found by eliminating x_2 from Eq. (3.65) by using Eq. (3.66), yielding

$$F_{ag} = \frac{(K_{se} + K_{lt})}{K_{se}} T_{ag} + K_{lt}x_1 . \tag{3.69}$$

At primary position, $x_{p1} + x_{p2} = -0.003$. To remove x_1 from Eq. (3.69), note that $x_1 = x - x_{p1}$ and $x_{p1} = -0.002167$, thus

$$F_{ag} = \frac{(K_{se} + K_{lt})}{K_{se}} T_{ag} + K_{lt} \left(x - x_{p1} \right) . \tag{3.70}$$

We next eliminate T_{ag} from Eq. (3.70) by using Eq. (3.58), yielding

$$F_{ag} = 0.14 + 0.0185\theta \text{ N} . \tag{3.71}$$

The static antagonist active state tension is found by eliminating x_3 from Eq. (3.67) by using Eq. (3.68), yielding

$$F_{ant} = \frac{(K_{se} + K_{lt})}{K_{se}} T_{ant} - K_{lt}x_4 . \tag{3.72}$$

To remove x_4 from Eq. (3.72), note that $x_4 = x + x_{p4}$ and $x_{p4} = 0.002167$, thus

$$F_{ant} = \frac{(K_{se} + K_{lt})}{K_{se}} T_{ant} - K_{lt} \left(x + x_{p4} \right) . \tag{3.73}$$

Next, we eliminate T_{ant} from Eq. (3.73) by using Eq. (3.59), yielding

$$F_{ant} = 0.14 - 0.0098\theta \text{ N} . \tag{3.74}$$

To satisfy Eq. (3.64) and ensure only positive active state tensions, Eqs. (3.71) and (3.74) are written with the following constraints.

$$F_{ag} = \begin{cases} 0.14 + 0.0185\theta & \text{N for } \theta < 14.23° \\ 0.0283\theta & \text{N for } \theta \geq 14.23° \end{cases}$$

$$F_{ant} = \begin{cases} 0.14 - 0.0098\theta & \text{N for } \theta < 14.23° \\ 0 & \text{N for } \theta \geq 14.23° . \end{cases} \tag{3.75}$$

These estimates of static active state tensions are in good agreement with those determined from Robinson's length tension innervation curves (Robinson, D., 1981).

Summarizing, the set of parameter estimates for the oculomotor plant are

$K_{se} = 125$ Nm^{-1}
$K_{lt} = 32$ Nm^{-1}
$K = 66.4$ Nm^{-1}
$B = 3.1$ Nsm^{-1}
$J = 2.2 \times 10^{-3}$ Ns^{2}m^{-1}
$B_{ag} = 3.4$ Nsm^{-1}
$B_{ant} = 1.2$ Nsm^{-1}
$\tau_{ac} = 0.009$ s
$\tau_{de} = 0.0054$ s
$\delta = 5.80288 \times 10^5$

The eigenvalues for the oculomotor plant using the parameter values above are -15, -66, -173, and $-1,293$.

DYNAMIC ACTIVE STATE TENSIONS

Initial estimates of the oculomotor control signals are based on an extrapolation of the eye movement trajectory. Specifically, the model is solved for the time to peak velocity and peak velocity to estimate the initial values for the control signal.

The saccadic eye movement model is solved via superposition using the classical technique. This is, separate solutions are found for the tensions operating between the time intervals 0 to t_1, and t_1 to ∞, and then combined to yield the complete solution. The forced response is given by

$$\theta_f(t) = \left(A_{11} + A_{21}e^{\frac{-t}{\tau_{ac}}} + A_{31}e^{\frac{-t}{\tau_{de}}} \right) u(t) + \left(A_{12} + A_{22}e^{\frac{-(t-t_1)}{\tau_{ac}}} + A_{32}e^{\frac{-(t-t_1)}{\tau_{de}}} \right) u(t-t_1) \qquad (3.76)$$

where

$$A_{11} = \frac{\delta (K_{se} + K_{lt}) F_p}{C_0}$$

$$D_{ac} = \frac{1}{\tau_{ac}^4} - \frac{C_3}{\tau_{ac}^3} + \frac{C_2}{\tau_{ac}^2} - \frac{C_1}{\tau_{ac}} + C_0$$

$$A_{21} = \frac{\delta \left((K_{se} + K_{lt}) - \frac{B_{ant}}{\tau_{ac}} \right) \left(F_{g0} - F_p \right)}{D_{ac}}$$

$$D_{ac} = \frac{1}{\tau_{de}^4} - \frac{C_3}{\tau_{de}^3} + \frac{C_2}{\tau_{de}^2} - \frac{C_1}{\tau_{de}} + C_0$$

$$A_{31} = \frac{\delta F_{t0} \left((K_{se} + K_{lt}) - \frac{B_{ag}}{\tau_{de}} \right) \left(F_{g0} - F_p \right)}{D_{de}}$$

$$A_{12} = \frac{-\delta \left(K_{se} + K_{lt} \right) \left(F_p - F_{gs} + F_{ts} \right)}{C_0}$$

$$A_{22} = \frac{-\delta \left(\left((K_{se} + K_{lt}) - \frac{B_{ag}}{\tau_{ac}} \right) \left(F_{t0} e^{\frac{-t_1}{\tau_{de}}} - F_{ts} \right) + \left((K_{se} + K_{lt}) - \frac{B_{ant}}{\tau_{ac}} \right) \left(e^{\frac{-t_1}{\tau_{de}}} \left(F_{g0} - F_p \right) \right) \right)}{D_{ac}}$$

$$A_{32} = \frac{\delta \left(F_{t0} e^{\frac{-t_1}{\tau_{de}}} \left((K_{se} + K_{lt}) - \frac{B_{ag}}{\tau_{de}} \right) + \left((K_{se} + K_{lt}) - \frac{B_{ant}}{\tau_{de}} \right) \left(F_p - F_{gs} + e^{\frac{-t_1}{\tau_{ac}}} \left(F_{g0} - F_p \right) \right) \right)}{D_{de}} .$$

The natural response is given by

$$\theta_n(t) = \left(K_{11} e^{a_1 t} + K_{21} e^{a_2 t} + K_{31} e^{a_3 t} + K_{41} e^{a_4 t} \right) u(t)$$
$$+ \left(K_{12} e^{a_1(t-t_1)} + K_{22} e^{a_2(t-t_1)} + K_{32} e^{a_3(t-t_1)} + K_{42} e^{a_4(t-t_1)} \right) u(t - t_1) \qquad (3.77)$$

where K_{ij} are the constants determined from the system's initial conditions, and a_i are the eigenvalues. The complete solution is

$$\theta(t) = \theta_n(t) + \theta_f(t)$$
$$= \left(K_{11} e^{a_1 t} + K_{21} e^{a_2 t} + K_{31} e^{a_3 t} + K_{41} e^{a_4 t} + A_{11} + A_{21} e^{\frac{-t}{\tau_{ac}}} + A_{31} e^{\frac{-t}{\tau_{de}}} \right) u(t)$$
$$+ \left(\begin{array}{c} K_{12} e^{a_1(t-t_1)} + K_{22} e^{a_2(t-t_1)} + K_{32} e^{a_3(t-t_1)} + K_{42} e^{a_4(t-t_1)} \\ + A_{12} + A_{22} e^{\frac{-(t-t_1)}{\tau_{ac}}} + A_{32} e^{\frac{-(t-t_1)}{\tau_{de}}} \end{array} \right) u(t - t_1) . \qquad (3.78)$$

The initial conditions are specified with the system at rest at primary position (looking straight ahead), that is, $\theta(0) = 0$. Note that Eq. (3.78) assumes the eye movement starts at time zero. In fact, the response to the target movement starts only after a variable latent period. This time delay is omitted for simplicity in writing the results and may be incorporated quite easily at a later time.

From the solution given in Eq. (3.78), the time to peak velocity and peak velocity are determined to yield estimates of agonist pulse magnitude and filter time constants. The time to peak velocity, t_{mv}, is calculated from

$$\left. \frac{\partial^2 \theta}{\partial t^2} \right|_{t=t_{mv}} = 0$$

or

$$0 = \left(a_1^2 K_{11} e^{a_1 t_{mv}} + a_2^2 K_{21} e^{a_2 t_{mv}} + a_3^2 K_{31} e^{a_3 t_{mv}} + a_4^2 K_{41} e^{a_4 t_{mv}} + \frac{A_{21}}{\tau_{ac}^2} e^{\frac{-t_{mv}}{\tau_{ac}}} + \frac{A_{31}}{\tau_{de}^2} e^{\frac{-t_{mv}}{\tau_{de}}} \right) u(t)$$
$$+ \left(\begin{array}{c} a_1^2 K_{12} e^{a_1 t_{mv}} e^{-a_1 t_1} + a_2^2 K_{22} e^{a_2 t_{mv}} e^{-a_2 t_1} + a_3^2 K_{32} e^{a_3 t_{mv}} e^{-a_3 t_1} + a_4^2 K_{42} e^{a_4 t_{mv}} e^{-a_4 t_1} \\ + \frac{A_{22}}{\tau_{ac}^2} e^{\frac{-t_{mv}}{\tau_{ac}}} e^{\frac{t_1}{\tau_{ac}}} + \frac{A_{32}}{\tau_{de}^2} e^{\frac{-t_{mv}}{\tau_{de}}} e^{\frac{t_1}{\tau_{de}}} \end{array} \right) u(t - t_1) . \qquad (3.79)$$

Equation (3.79) is iteratively solved for t_{mv} using a first-order Taylor series approximation for the exponential terms involving t_{mv}.

$$e^{a_i t_{mv}^{j+1}} = e^{a_i t_{mv}^{j}} e^{a_i t_{mv}^{j+1} - t_{mv}^{j}} \approx e^{a_i t_{mv}^{j}} \left(1 + a_i \left(t_{mv}^{j+1} - t_{mv}^{j} \right) \right) . \tag{3.80}$$

Substitute Eq. (3.80) into Eq. (3.79), yielding

$$0 = \left(\begin{array}{l} a_1^2 K_{11} e^{a_1 t_{mv}^{j}} \left(1 + a_1 \left(t_{mv}^{j+1} - t_{mv}^{j} \right) \right) + a_2^2 K_{21} e^{a_2 t_{mv}^{j}} \left(1 + a_2 \left(t_{mv}^{j+1} - t_{mv}^{j} \right) \right) \\ + a_3^2 K_{31} e^{a_3 t_{mv}^{j}} \left(1 + a_3 \left(t_{mv}^{j+1} - t_{mv}^{j} \right) \right) + a_4^2 K_{41} e^{a_4 t_{mv}^{j}} \left(1 + a_4 \left(t_{mv}^{j+1} - t_{mv}^{j} \right) \right) \\ + \frac{A_{21}}{\tau_{ac}^2} e^{\frac{t_{mv}^{j}}{\tau_{ac}}} \left(1 + \frac{\left(t_{mv}^{j+1} - t_{mv}^{j} \right)}{\tau_{ac}} \right) + \frac{A_{31}}{\tau_{de}^2} e^{\frac{t_{mv}^{j}}{\tau_{de}}} \left(1 + \frac{\left(t_{mv}^{j+1} - t_{mv}^{j} \right)}{\tau_{de}} \right) \end{array} \right) u(t)$$

$$+ \left(\begin{array}{l} a_1^2 K_{12} e^{a_1 t_{mv}^{j}} \left(1 + a_1 \left(t_{mv}^{j+1} - t_{mv}^{j} \right) \right) e^{-a_1 t_1} + a_2^2 K_{22} e^{a_2 t_{mv}^{j}} \left(1 + a_2 \left(t_{mv}^{j+1} - t_{mv}^{j} \right) \right) e^{-a_2 t_1} \\ + a_3^2 K_{32} e^{a_3 t_{mv}^{j}} \left(1 + a_3 \left(t_{mv}^{j+1} - t_{mv}^{j} \right) \right) e^{-a_3 t_1} + a_4^2 K_{42} e^{a_4 t_{mv}^{j}} \left(1 + a_4 \left(t_{mv}^{j+1} - t_{mv}^{j} \right) \right) e^{-a_4 t_1} \\ + \frac{A_{22}}{\tau_{ac}^2} e^{\frac{t_{mv}^{j}}{\tau_{ac}}} \left(1 + \frac{\left(t_{mv}^{j+1} - t_{mv}^{j} \right)}{\tau_{ac}} \right) e^{\frac{t_1}{\tau_{ac}}} + \frac{A_{32}}{\tau_{de}^2} e^{\frac{t_{mv}^{j}}{\tau_{de}}} \left(1 + \frac{\left(t_{mv}^{j+1} - t_{mv}^{j} \right)}{\tau_{de}} \right) e^{\frac{t_1}{\tau_{de}}} \end{array} \right) u(t - t_1) .$$

Separating like terms gives

$$0 = \left(\begin{array}{l} a_1^2 K_{11} e^{a_1 t_{mv}^{j}} \left(1 - a_1 t_{mv}^{j} \right) + a_2^2 K_{21} e^{a_2 t_{mv}^{j}} \left(1 - a_2 t_{mv}^{j} \right) \\ + a_3^2 K_{31} e^{a_3 t_{mv}^{j}} \left(1 - a_3 t_{mv}^{j} \right) + a_4^2 K_{41} e^{a_4 t_{mv}^{j}} \left(1 - a_4 t_{mv}^{j} \right) \\ + \frac{A_{21}}{\tau_{ac}^2} e^{\frac{t_{mv}^{j}}{\tau_{ac}}} \left(1 - \frac{t_{mv}^{j}}{\tau_{ac}} \right) + \frac{A_{31}}{\tau_{de}^2} e^{\frac{t_{mv}^{j}}{\tau_{de}}} \left(1 - \frac{t_{mv}^{j}}{\tau_{de}} \right) \end{array} \right) u(t)$$

$$+ t_{mv}^{j+1} \left(\begin{array}{l} a_1^3 K_{11} e^{a_1 t_{mv}^{j}} + a_2^3 K_{21} e^{a_2 t_{mv}^{j}} \\ + a_3^3 K_{31} e^{a_3 t_{mv}^{j}} + a_4^3 K_{41} e^{a_4 t_{mv}^{j}} \\ + \frac{A_{21}}{\tau_{ac}^3} e^{\frac{t_{mv}^{j}}{\tau_{ac}}} + \frac{A_{31}}{\tau_{de}^3} e^{\frac{t_{mv}^{j}}{\tau_{de}}} \end{array} \right) u(t)$$

$$+ \begin{pmatrix} a_1^2 K_{12} e^{a_1 t_{mv}^j} \left(1 - a_1 t_{mv}^j\right) e^{-a_1 t_1} + a_2^2 K_{22} e^{a_2 t_{mv}^j} \left(1 - a_2 t_{mv}^j\right) e^{-a_2 t_1} \\ + a_3^2 K_{32} e^{a_3 t_{mv}^j} \left(1 - a_3 t_{mv}^j\right) e^{-a_3 t_1} + a_4^2 K_{42} e^{a_4 t_{mv}^j} \left(1 - a_4 t_{mv}^j\right) e^{-a_4 t_1} \\ + \frac{A_{22}}{\tau_{ac}^2} e^{\frac{t_{mv}^j}{\tau_{ac}}} \left(1 - \frac{t_{mv}^j}{\tau_{ac}}\right) e^{\frac{t_1}{\tau_{ac}}} + \frac{A_{32}}{\tau_{de}^2} e^{\frac{t_{mv}^j}{\tau_{de}}} \left(1 - \frac{t_{mv}^j}{\tau_{de}}\right) e^{\frac{t_1}{\tau_{de}}} \end{pmatrix} u(t - t_1)$$

$$+ t_{mv}^{j+1} \begin{pmatrix} a_1^3 K_{12} e^{a_1 t_{mv}^j} e^{-a_1 t_1} + a_2^3 K_{22} e^{a_2 t_{mv}^j} e^{-a_2 t_1} + a_3^3 K_{32} e^{a_3 t_{mv}^j} e^{-a_3 t_1} \\ + a_4^3 K_{42} e^{a_4 t_{mv}^j} e^{-a_4 t_1} + \frac{A_{22}}{\tau_{ac}^3} e^{\frac{t_{mv}^j}{\tau_{ac}}} e^{\frac{t_1}{\tau_{ac}}} + \frac{A_{32}}{\tau_{de}^3} e^{\frac{t_{mv}^j}{\tau_{de}}} e^{\frac{t_1}{\tau_{de}}} \end{pmatrix} u(t - t_1) .$$

Next, we solve for t_{mv}^{j+1}

$$t_{mv}^{j+1} = \frac{\begin{pmatrix} a_1^2 K_{11} e^{a_1 t_{mv}^j} \left(1 - a_1 t_{mv}^j\right) \\ + a_2^2 K_{21} e^{a_2 t_{mv}^j} \left(1 - a_2 t_{mv}^j\right) \\ + a_3^2 K_{31} e^{a_3 t_{mv}^j} \left(1 - a_3 t_{mv}^j\right) \\ + a_4^2 K_{41} e^{a_4 t_{mv}^j} \left(1 - a_4 t_{mv}^j\right) \\ + \frac{A_{21}}{\tau_{ac}^2} e^{\frac{t_{mv}^j}{\tau_{ac}}} \left(1 - \frac{t_{mv}^j}{\tau_{ac}}\right) \\ + \frac{A_{31}}{\tau_{de}^2} e^{\frac{t_{mv}^j}{\tau_{de}}} \left(1 - \frac{t_{mv}^j}{\tau_{de}}\right) \end{pmatrix} u(t) + \begin{pmatrix} a_1^2 K_{12} e^{a_1 t_{mv}^j} \left(1 - a_1 t_{mv}^j\right) e^{-a_1 t_1} \\ + a_2^2 K_{22} e^{a_2 t_{mv}^j} \left(1 - a_2 t_{mv}^j\right) e^{-a_2 t_1} \\ + a_3^2 K_{32} e^{a_3 t_{mv}^j} \left(1 - a_3 t_{mv}^j\right) e^{-a_3 t_1} \\ + a_4^2 K_{42} e^{a_4 t_{mv}^j} \left(1 - a_4 t_{mv}^j\right) e^{-a_4 t_1} \\ + \frac{A_{22}}{\tau_{ac}^2} e^{\frac{t_{mv}^j}{\tau_{ac}}} \left(1 - \frac{t_{mv}^j}{\tau_{ac}}\right) e^{\frac{t_1}{\tau_{ac}}} \\ + \frac{A_{32}}{\tau_{de}^2} e^{\frac{t_{mv}^j}{\tau_{de}}} \left(1 - \frac{t_{mv}^j}{\tau_{de}}\right) e^{\frac{t_1}{\tau_{de}}} \end{pmatrix} u(t - t_1)}{} \qquad (3.81)$$

$$t_{mv}^{j+1} \begin{pmatrix} a_1^3 K_{11} e^{a_1 t_{mv}^j} + a_2^3 K_{21} e^{a_2 t_{mv}^j} \\ + a_3^3 K_{31} e^{a_3 t_{mv}^j} + a_4^3 K_{41} e^{a_4 t_{mv}^j} \\ + \frac{A_{21}}{\tau_{ac}^3} e^{\frac{t_{mv}^j}{\tau_{ac}}} + \frac{A_{31}}{\tau_{de}^3} e^{\frac{t_{mv}^j}{\tau_{de}}} \end{pmatrix} u(t)$$

$$+ \begin{pmatrix} a_1^3 K_{12} e^{a_1 t_{mv}^j} e^{-a_1 t_1} + a_2^3 K_{22} e^{a_2 t_{mv}^j} e^{-a_2 t_1} \\ + a_3^3 K_{32} e^{a_3 t_{mv}^j} e^{-a_3 t_1} + a_4^3 K_{42} e^{a_4 t_{mv}^j} e^{-a_4 t_1} \\ + \frac{A_{22}}{\tau_{ac}^3} e^{\frac{t_{mv}^j}{\tau_{ac}}} e^{\frac{t_1}{\tau_{ac}}} + \frac{A_{32}}{\tau_{de}^3} e^{\frac{t_{mv}^j}{\tau_{de}}} e^{\frac{t_1}{\tau_{de}}} \end{pmatrix} u(t - t_1)$$

where the $(j + 1)^{th}$ iterates of t_{mv}^{j+1} are calculated from the jth iterates of t_{mv}^j. The procedure of solving for time at peak velocity begins with specifying an initial guess, t_{mv}^0, and using Eq. (3.81) to solve for t_{mv}^{j+1}, and iterating until the desired degree of accuracy is achieved. Clearly, the solution for t_{mv} involves two distinct cases: case 1 with t_{mv} dependent on the size of the saccade and case 2 with t_{mv} independent of the size of the saccade. Case 1 involves saccades in which the agonist pulse duration is in the interval $[0, t_1^c]$, where t_1^c is the maximum value of t_1 in which t_{mv} is a function of saccade displacement. The case 1 solution for t_{mv} yields a value greater than t_1 due to the mechanical components of the oculomotor model. The case 2 solution of t_{mv} yields $t_{mv} = t_1^c$ for all saccades when $t_1 > t_1^c$. These results agree with the characteristics of the main sequence diagram.

Peak velocity is calculated from

$$
\begin{aligned}
\dot{\theta}(t_{mv}) = {} & \left(a_1 K_{11} e^{a_1 t_{mv}} + a_2 K_{21} e^{a_2 t_{mv}} + a_3 K_{31} e^{a_3 t_{mv}} + a_4 K_{41} e^{a_4 t_{mv}} - \frac{A_{21}}{\tau_{ac}} e^{\frac{-t_{mv}}{\tau_{ac}}} - \frac{A_{31}}{\tau_{de}} e^{\frac{-t_{mv}}{\tau_{de}}} \right) u(t) \\
& + \left(\begin{array}{l} a_1 K_{12} e^{a_1 t_{mv}} e^{-a_1 t_1} + a_2 K_{22} e^{a_2 t_{mv}} e^{-a_1 t_1} + a_3 K_{32} e^{a_3 t_{mv}} e^{-a_1 t_1} + a_4 K_{42} e^{a_4 t_{mv}} e^{-a_1 t_1} \\ - \frac{A_{22}}{\tau_{ac}} e^{\frac{-t_{mv}}{\tau_{ac}}} e^{-a_1 t_1} - \frac{A_{32}}{\tau_{de}} e^{\frac{-t_{mv}}{\tau_{de}}} e^{-a_1 t_1} \end{array} \right) u(t - t_1).
\end{aligned}
$$

$$(3.82)$$

The diagrams in Fig. 3.34 illustrate the effect of agonist pulse magnitude, pulse duration, and activation time constant on peak velocity using the parameter estimates described in this section.

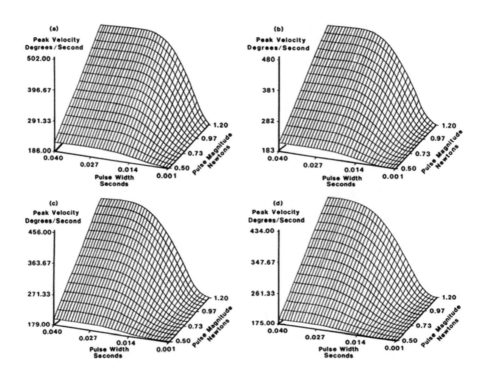

Figure 3.34: Diagram illustrating the effect of the agonist pulse magnitude, pulse duration and activation time constant on peak velocity. (a) $\tau_{ac} = 0.004$ s; (b) $\tau_{ac} = 0.007$ s; (c) $\tau_{ac} = 0.010$ s; (d) $\tau_{ac} = 0.013$ s; $\tau_{de} = 0.005$ s for all saccades.

As expected, the pulse magnitude strongly affects peak velocity, increasing F_p, increases $\dot{\theta}$ (t_{mv}) for all pulse durations. Additionally, increasing F_p does not affect t_1^c; that is, t_1^c is independent of the pulse magnitude. Increasing the pulse duration from zero increases peak velocity until $t_{mv} > t_1^c$ as computed from Eq. (3.82), after which peak velocity remains a constant as pulse duration increases. Increasing the activation time constant increases t_1^c, while slightly reducing the peak velocity $\dot{\theta}$ (t_{mv}). It should be noted that the effects of the deactivation time constant during the pulse phase of the trajectory does not significantly influence the time to peak velocity or the peak velocity.

Saccadic eye movement data and physiological conditions determine the initial estimates for the dynamic active state tensions during a saccade. First, the saccadic eye movement data is analyzed using the two-point central difference method from which estimates of velocity and acceleration are computed (the two-point central difference method is described in Chapter 4). Velocity estimates are computed with a step size of 3 and acceleration estimates are computed with a step size of 4. Values for the time to peak velocity, t_{mv}, and peak velocity, $\dot{\theta}$ (t_{mv}), are calculated from the data, and used to determine F_p, t_1 and τ_{ac} based on the theoretical peak velocity investigation presented here and summarized in the graphs in Fig. 3.34. The initial estimate of the deactivation time constant is also estimated directly from the time interval from t_1 to the end of the saccade, divided by four (i.e., within four time constants an exponential reaches steady state).

SYSTEM IDENTIFICATION TECHNIQUE AND ESTIMATION RESULTS

Parameter estimates and inputs for the model of the oculomotor system are found using the system identification technique as previously described. Final estimates for all parameters and inputs were determined with a conjugate gradient search program, initialized using published physiological data and experimental data. Figure 3.35 shows estimation routine results for a 15° target displacement in the time and frequency domain. The accuracy of these results is typical for all target displacements with all subjects tested, except for saccades with glissadic or dynamic overshoot (saccades with glissadic or dynamic overshoot are discussed in Chapter 6). Figures 3.36 and 3.37 further illustrates the accuracy of the system identification technique parameter estimation routine by the close agreement of the velocity and acceleration estimates with the two-point central differences estimates.

Displayed in Fig. 3.38 are the system identification technique estimates of agonist pulse magnitude as a function of displacement for the three subjects tested. The estimated agonist pulse magnitude showed more variation within each target movement than between target movements. One pronounced feature evident from this graph is the apparent lack of a strong relationship between agonist pulse magnitude and displacement related by other investigators. In fact, pulse magnitude is evidently independent of the size of the target displacement. These results are consistent with a control of saccadic eye movements that is time-optimal as described in the next section. The magnitude of the agonist pulse is a maximum regardless of the size of the saccade and only the duration of the agonist pulse affects the size of the saccade.

Analogous to the agonist pulse magnitude relationship with saccade magnitude, the activation and deactivation time constants showed more variation within each target movement than between

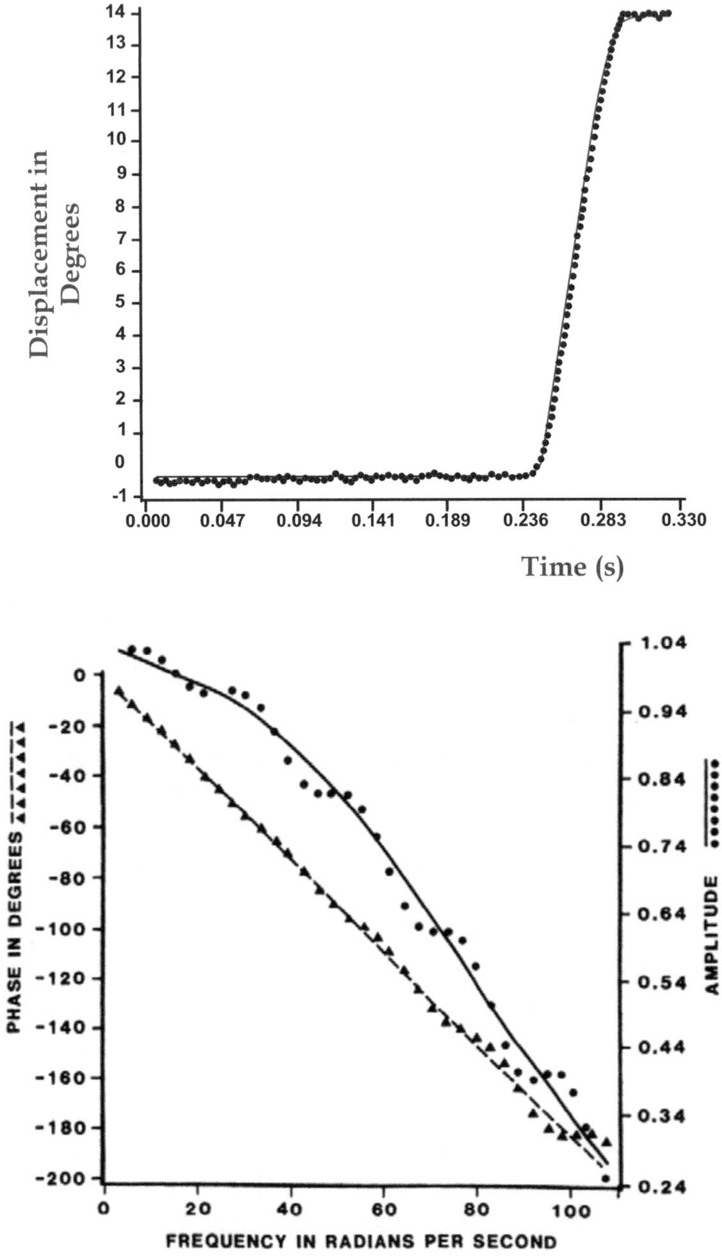

Figure 3.35: Time and frequency response for a 15 degree saccadic eye movement. Solid and dashed lines are the predictions of the saccadic eye movement model with final parameter estimates computed using the system identification technique. Dots and triangles are the data.

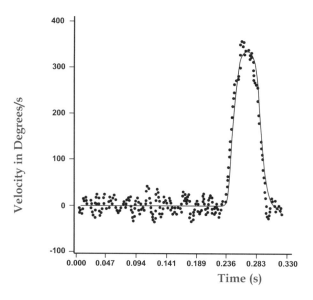

Figure 3.36: A plot of the velocity estimates (dots) and velocity simulation results (solid line) from the modified linear homeomorphic model.

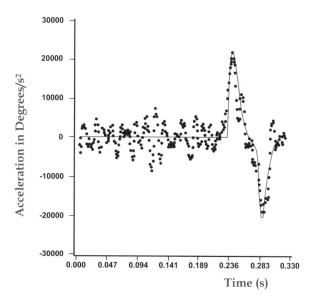

Figure 3.37: A plot of the acceleration estimates (dots) and acceleration simulation results (solid line) from the modified linear homeomorphic model.

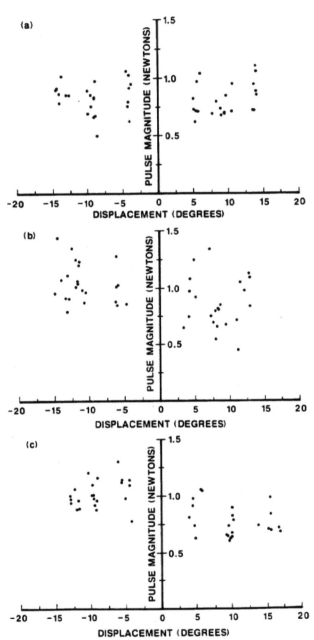

Figure 3.38: System identification technique estimate of the agonist pulse magnitude as a function of saccade magnitude for the three subjects tested (a), (b), and (c). Figures 3.35–3.37 correspond to subject (a).

target movements for this subject as displayed in Fig. 3.39. The average activation time constant is

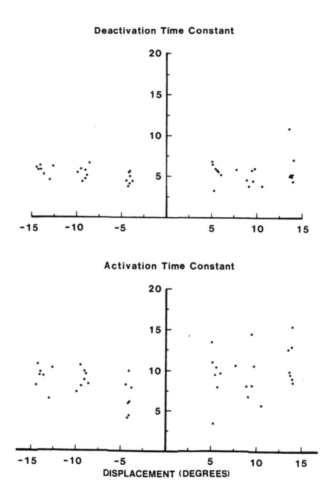

Figure 3.39: System identification technique estimates of the activation and deactivation time constant as a function of saccade magnitude.

9.0 ms, with a range of values from 3.7 to 15.7 ms. The average deactivation time constant is 5.4 ms, with a range of values from 3.5 ms to 7.2 ms (one outlier with a value of 11.1 falls outside this range). No trends are noted in the data between either time constant and saccade magnitude.

3.6.3   TIME OPTIMAL CONTROL OF SACCADIC EYE MOVEMENTS

Saccadic eye movements, among the fastest voluntary muscle movements the human body is capable of producing, are characterized by a rapid shift of gaze from one point of fixation to another. Although the purpose for such an eye movement is obvious, that is, to quickly redirect the eyeball to the target, the neuronal control strategy is not. For instance, does the word "quickly" in the previous sentence imply the most rapid movement possible, or simply a fast as opposed to a slow movement? To reach a destination in minimum time, the input to the oculomotor system must be bang-bang according to Pontryagin's minimum principle; that is, the oculomotor system is either maximally or minimally stimulated during the saccadic eye movement. With this control strategy, saccade magnitude is affected only by the length of the time intervals during which the system is maximally or minimally stimulated. This section describes an investigation utilizing Pontryagin's minimum principle and system identification techniques to estimate muscle active state tensions during horizontal saccadic eye movements in order to better understand the neuronal control strategy.

To detail the neuronal control strategy, it is necessary to understand the effect of the saccadic innervation signals on the oculomotor plant and the resultant response. Many investigators have extensively studied the saccadic innervation signals, which are described by pulse-step waveforms. At the start of a saccade, the agonist muscle is strongly stimulated, and the antagonist muscle is completely inhibited. After a brief time interval, this is followed by a decrease in agonist stimulation and an increase in antagonist stimulation to tonic levels necessary to maintain the eyeball in its new position. Collins states that the amplitude and duration of the saccadic innervation signal determines the magnitude of each saccade (Collins et al., 1975). Specifically, he determined a logarithmic relationship between innervation amplitude and saccade magnitude. Zee and his co-workers assumed that a local feedback loop automatically controlled the amplitude and duration of the saccadic innervation signal (Zee et al., 1976). The difference between the internal representation of the present eye position and the desired eye position determines the saccadic innervation signal. Note that Zee et al. still hypothesize a pulse-step innervation signal but base the pulse size and duration on a nonlinear velocity function. Bahill reports a nonlinear relationship between saccade innervation magnitude and saccade magnitude, and a linear relationship between saccade innervation duration and saccade magnitude (Bahill, A., 1980). If these authors are correct in their assertion that saccadic magnitude is a function of both the duration and the amplitude of the innervation signal, and not duration alone, then neuronal control does not operate with a bang-bang or minimum-time strategy.

While investigators have recorded the innervation signal from several types of motoneurons that drive the eyeball during a saccade, they have not directly measured the active state tensions responsible for this movement. Collins and his co-workers have measured the muscle tension in vivo at the muscle tendon during unrestrained human eye movement using a miniature "C" gauge force transducer (Collins et al., 1975). The active state tension, however, is distributed throughout the muscle and cannot be directly measured since it is modified by the viscoelasticity of the muscle. Very little is known about the dynamic active state tensions generated in the antagonist-agonist muscle pair during a saccade and their relationship to the saccadic innervation signal. Naturally, under static conditions during fixation, the active state tensions are proportional to the innervation signal, and a constant firing frequency produces a constant active state tension. During a saccade, the agonist active state tension changes rapidly, rising in a matter of milliseconds to a new level approximately tenfold higher than during fixation, and then falling to the new fixation level. The proportional relationship that exists for innervation and active state tension during fixation is not valid during a saccade due to the effect of saturation and filtering of the input signal. Thus, the exact shape of the input to the muscle is uncertain. The active state tensions are typically modeled by low-pass filtering the innervation signal (1981). While very little is known about the activation and deactivation time constants due to lack of in vivo testing, Bahill has estimated their values to be between 0.2 and 13 ms based on the rise of the isometric force during electrical stimulation (Bahill, A., 1981). In the previous section, we estimated the average activation time constant is 9.0 ms, with a range of values from 3.7 to 15.7 ms. The average deactivation time constant is 5.4 ms, with a range of values from 3.5 ms to 7.2 ms.

Robinson presented data that seem to contradict the previous relationship of amplitude and duration with saccade magnitude (Robinson, D., 1981). From Robinson, D. (1981, Fig. 4), note that the agonist motoneuron burst peaks at the same amplitude and then drops to a constant level during the saccade regardless of the size of the retinal error. Since the motoneurons fire well above 200 Hz for the initial pulse phase of the trajectory regardless of the amplitude of the saccade, only the duration of the agonist pulse is a function of the saccade displacement. Under these conditions, the eyeball appears to be driven to its destination in minimum- time for saccades of all sizes.

Other researchers have investigated the neuronal control strategy of saccadic eye movements using optimal control theory. While these investigators concluded that each saccadic eye movement is driven to achieve final eye position in minimum-time, they reported different neuronal control strategies. Clark and Stark, 1975 postulated second-order time-optimal control signals, but they observed a first-order time-optimal control simulation solution when using the same model in both analyses (Clark and Stark, 1975). While never completely rectifying these differences, Clark and Stark concluded that the saccadic eye movement neuronal control strategy is first-order time-optimal. Clark and Stark did not give switch-time details or comment on the pulse magnitude-saccade magnitude relationship. Lehman and Stark, however, reported a second-order time-optimal controller using a simplified saccadic eye movement model which excluded the activation and deactivation time constants (Lehman and Stark, 1979). Further, Lehman and Stark indicated that they were unable to solve the optimality problem using a saccadic eye movement model with activation and deactivation time constants. Lehman and Stark also reported agonist pulse magnitude as a function of saccade amplitude, which violates the bang-bang controller.

This section presents an optimal control investigation of horizontal saccadic eye movements based on Pontryagin's minimum principle with a linear oculomotor model in which activation and deactivation time constants are explicitly included. Based on the optimality solution, it is shown that horizontal saccadic eye movement neuronal control is a first-order time-optimal control signal. The concepts underlying this hypothesis are

1) the agonist pulse is maximum regardless of the amplitude of the saccade

2) only the duration of the agonist pulse effects the size of the saccade

The antagonist muscle is assumed to be completely inhibited during the period of maximum stimulation for the agonist muscle. Furthermore, higher order signals are found not to be time-optimal. A quantitative analysis of saccadic eye movement data is also presented to support the hypothesis that the saccadic neuronal control mechanism operates to achieve final eye position in minimum-time under a first-order controller. Thus, a consistent first-order time-optimal neuronal control strategy is demonstrated which also agrees with experimental data analysis.

TIME-OPTIMAL NEURONAL CONTROL STRATEGY

The hypothesis that the eyeball is driven to its destination in minimum-time for saccades of all sizes is investigated using optimal control theory based on the minimum principle of Pontryagin with a linear oculomotor model. In order to state the optimal control problem to be solved, we describe the model in terms of state variables with the agonist and antagonist active state tensions explicitly included as follows.

$$
\begin{aligned}
\dot{\theta}_1 &= \theta_2 \\
\dot{\theta}_2 &= \theta_3 \\
\dot{\theta}_3 &= \theta_4 \\
\dot{\theta}_4 &= C_0\theta_1 - C_1\theta_2 - C_2\theta_3 - C_3\theta_4 \\
&\quad + \delta\left((K_{se} + K_{lt})(\theta_5 - \theta_6) + B_{ant}\frac{n_1 - \theta_5}{\tau_{ag}} - B_{ag}\frac{n_2 - \theta_6}{\tau_{ant}}\right) \\
\dot{\theta}_5 &= \frac{n_1 - \theta_5}{\tau_{ag}} \\
\dot{\theta}_6 &= \frac{n_2 - \theta_6}{\tau_{ant}}
\end{aligned}
\tag{3.83}
$$

where

$\theta_1 = \theta = $ angular position
$\theta_2 = \dot{\theta}_1 = $ angular velocity
$\theta_3 = \dot{\theta}_2 = $ angular acceleration
$\theta_4 = \dot{\theta}_3 = $ angular jerk
$\theta_5 = $ agonist active state tension
$\theta_6 = $ antagonist active state tension
$n_1 = $ agonist neurological controller
$n_2 = $ antagonist neurological controller
$C_i = $ mechanical components of the oculomotor plant

Note that the filter time constants τ_{ag} and τ_{ant} are functions of time, that is

$$
\begin{aligned}
\tau_{ag} &= \tau_{ac}(u(t) - u(t - t_1)) + \tau_{de}u(t_1) \\
\tau_{ant} &= \tau_{de}(u(t) - u(t - t_1)) + \tau_{ac}u(t_1).
\end{aligned}
\tag{3.84}
$$

Due to physiological constraints, the agonist and antagonist neurological controller must satisfy $0 \leq n_i \leq n_{max}$. Equation (3.83) is written in matrix form as

$$\dot{\theta} = \mathbf{A}\theta + \mathbf{B}\mathbf{n} \tag{3.85}$$

where

$$\theta = \begin{bmatrix} \theta_1 \\ \theta_2 \\ \theta_3 \\ \theta_4 \\ \theta_5 \\ \theta_6 \end{bmatrix} \quad \mathbf{A} = \begin{bmatrix} 0 & 1 & 0 & 0 & 0 & 0 \\ 0 & 0 & 1 & 0 & 0 & 0 \\ 0 & 0 & 0 & 1 & 0 & 0 \\ -C_0 & -C_1 & -C_2 & -C_3 & \delta\left((\dot{K_{se}} + K_{lt}) - \frac{B_{ant}}{\tau_{ag}}\right) & -\delta\left((K_{se} + K_{lt}) - \frac{B_{ag}}{\tau_{ant}}\right) \\ 0 & 0 & 0 & 0 & \frac{-1}{\tau_{ag}} & 0 \\ 0 & 0 & 0 & 0 & 0 & \frac{-1}{\tau_{ant}} \end{bmatrix}$$

$$\mathbf{B} = \begin{bmatrix} 0 & 0 \\ 0 & 0 \\ 0 & 0 \\ \frac{\delta B_{ant}}{\tau_{ag}} & -\frac{\delta B_{ag}}{\tau_{ant}} \\ \frac{1}{\tau_{ag}} & 0 \\ 0 & \frac{1}{\tau_{ant}} \end{bmatrix} \quad \mathbf{n} = \begin{bmatrix} n_1 \\ n_2 \end{bmatrix}.$$

It should be clear that the system matrix $\mathbf{A}$ is stepwise time-varying due to the time constants. The hypothesized time-optimal neural control strategy is to choose the control $\mathbf{n}(t)$ to transfer $\theta(0)$ according to $\dot{\theta} = \mathbf{A}\theta + \mathbf{B}\mathbf{n}$ to the destination D so that the functional

$$J(\mathbf{n}) = \int_0^{t_f} dt + \sum_1^6 \mathbf{G}\left(\theta_i\left(t_f\right) - D_i\right)^2 \tag{3.86}$$

is minimized, where the terminal time t_f and $\theta(t_f)$ are unspecified. $\mathbf{G}$ is a weighting vector which determines the nearness of $\theta(t_f)$ to $\mathbf{D}$.

TIME-OPTIMAL CONTROL SOLUTION

The time-optimal control solution is investigated using the minimum principle of Pontryagin. The minimum principle states that the time-optimal input to the model must be bang-bang. Using the standard approach to the optimality problem, the gradient or steepest descent method, it is impossible to solve for the switch-time for the sixth order oculomotor model. Instead of simplifying the model in order to determine the optimal control, as did Lehman and Stark (1979), we decided to solve for the optimal switch-times directly.

Direct Optimal Switch-Time Evaluation   The difficulties associated with the gradient method are avoided by directly evaluating the optimal switch-time for the minimum-time controller based on the works of Pierre, D. (1969), Lee, E. (1960), and Smith, F. (1961). First, using Pontryagin's minimum principle, this system's minimum-time controller is of the bang-bang type. Due to physiological considerations, the agonist neurological control is fully stimulated at the start of the saccade and the antagonist neurological control is completely inhibited. At the switch-time, the controllers exchange values. Thus, all that is necessary to solve this problem is to specify the minimum-time controller switch-time. Since the oculomotor system is a time-varying system, the direct evaluation procedure in solving for the optimal switch-time by Pierre, D. (1969), Lee, E. (1960), and Smith, F. (1961) is not suitable since the commutativity condition is not satisfied (Kinariwala, B., 1961). Fortunately, the differential equation that describes saccadic eye movements with a bang-bang controller is readily solved using classical techniques (see Eq. (3.78), expanded below).

$$\theta(t) = \left(K_{11}e^{a_1 t} + K_{21}e^{a_2 t} + K_{31}e^{a_3 t} + K_{41}e^{a_4 t} + A_{11} + A_{21}e^{\frac{-t}{\tau_{ac}}} + A_{31}e^{\frac{-t}{\tau_{de}}} \right) u(t)$$
$$+ \left(\begin{matrix} K_{12}e^{a_1(t-t_1)} + K_{22}e^{a_2(t-t_1)} + K_{32}e^{a_3(t-t_1)} + K_{42}e^{a_4(t-t_1)} \\ + A_{12} + A_{22}e^{\frac{-(t-t_1)}{\tau_{ac}}} + A_{32}e^{\frac{-(t-t_1)}{\tau_{de}}} \end{matrix} \right) u(t-t_1) . \quad (3.87)$$

Thus, by avoiding direct evaluation of the state transition matrix, the direct evaluation computational procedure can be modified appropriately with the saccadic eye movement model solution to yield the optimal switch-time.

The saccadic eye movement model is solved via superposition by incorporating the bang-bang neurological controllers directly in the agonist and antagonist active state tensions and treating the active state tensions as inputs. That is, separate solutions are found for the tensions operating between the time intervals 0 to t_1 and $t > t_1$, and then combined to yield the complete solution.

The initial conditions are specified with the system at rest at primary position (looking straight ahead), that is, $\theta(0) = 0$. In general, the minimum-time controller is specified by selecting t_1 and t_n so that $t_f > t_n$ is minimum and $\theta(t_f) = \mathbf{D}$. At $\theta(t_f)$ we have

$$\theta(t_f) = K_{11}e^{a_1 t_f} + K_{21}e^{a_2 t_f} + K_{31}e^{a_3 t_f} + K_{41}e^{a_4 t_f} + A_{11} + A_{21}e^{\frac{-t_f}{\tau_{ac}}} + A_{31}e^{\frac{-t_f}{\tau_{de}}}$$
$$+ K_{12}e^{a_1(t_f-t_1)} + K_{22}e^{a_2(t_f-t_1)} + K_{32}e^{a_3(t_f-t_1)} + K_{42}e^{a_4(t_f-t_1)}$$
$$+ A_{12} + A_{22}e^{\frac{-(t_f-t_1)}{\tau_{ac}}} + A_{32}e^{\frac{-(t_f-t_1)}{\tau_{de}}} . \quad (3.88)$$

The only unknowns in Eq. (3.88) are t_1 and t_f. As before, Eq. (3.88) is solved for t_1 by using a first-order exponential Taylor series approximation iterative linearization technique as described previously. First, assume that t_f is known. Next, substitute the truncated exponential Taylor series approximation

$$e^{a_i t_1^{j+1}} = e^{a_i t_1^j} e^{a_i t_1^{j+1} - t_1^j} \approx e^{a_i t_1^j} \left(1 + a_i \left(t_1^{j+1} - t_1^j \right) \right) \quad (3.89)$$

into Eq. (3.88) for $e^{a_i t_1}$, which yields

$$
\begin{aligned}
\theta(t_f) &= K_{11}e^{a_1 t_f} + K_{21}e^{a_2 t_f} + K_{31}e^{a_3 t_f} + K_{41}e^{a_4 t_f} + A_{11} + A_{21}e^{\frac{-t_f}{\tau_{ac}}} + A_{31}e^{\frac{-t_f}{\tau_{de}}} \\
&+ K_{12}e^{a_1\left(t_f - t_1^j\right)}\left(a_1\left(t_1^{j+1} - t_1^j\right) + 1\right) + K_{22}e^{a_2\left(t_f - t_1^j\right)}\left(a_2\left(t_1^{j+1} - t_1^j\right) + 1\right) \\
&+ K_{32}e^{a_3\left(t_f - t_1^j\right)}\left(a_3\left(t_1^{j+1} - t_1^j\right) + 1\right) + K_{42}e^{a_4\left(t_f - t_1^j\right)}\left(a_4\left(t_1^{j+1} - t_1^j\right) + 1\right) \\
&+ A_{12} + A_{22}e^{\frac{-\left(t_f - t_1^j\right)}{\tau_{ac}}}\left(\frac{\left(t_1^{j+1} - t_1^j\right)}{\tau_{ac}} + 1\right) + A_{32}e^{\frac{-\left(t_f - t_1^j\right)}{\tau_{de}}}\left(\frac{\left(t_1^{j+1} - t_1^j\right)}{\tau_{de}} + 1\right).
\end{aligned}
\tag{3.90}
$$

Next, factor out the term t_1^{j+1} in Eq. (3.90), giving

$$
\begin{aligned}
\theta(t_f) &= K_{11}e^{a_1 t_f} + K_{21}e^{a_2 t_f} + K_{31}e^{a_3 t_f} + K_{41}e^{a_4 t_f} + A_{11} + A_{21}e^{\frac{-t_f}{\tau_{ac}}} + A_{31}e^{\frac{-t_f}{\tau_{de}}} \\
&+ K_{12}e^{a_1\left(t_f - t_1^j\right)}\left(1 - t_1^j a_1\right) + K_{22}e^{a_2\left(t_f - t_1^j\right)}\left(1 - a_2 t_1^j\right) + K_{32}e^{a_3\left(t_f - t_1^j\right)}\left(1 - t_1^j a_3\right) \\
&+ K_{42}e^{a_4\left(t_f - t_1^j\right)}\left(1 - a_4 t_1^j\right) + A_{12} + A_{22}e^{\frac{-\left(t_f - t_1^j\right)}{\tau_{ac}}}\left(1 - \frac{t_1^j}{\tau_{ac}}\right) + A_{32}e^{\frac{-\left(t_f - t_1^j\right)}{\tau_{de}}}\left(1 - \frac{t_1^j}{\tau_{de}}\right) \\
&+ t_1^{j+1}\left(\begin{array}{c} a_1 K_{12}e^{a_1\left(t_f - t_1^j\right)} + a_2 K_{22}e^{a_2\left(t_f - t_1^j\right)} + a_3 K_{32}e^{a_3\left(t_f - t_1^j\right)} + a_4 K_{42}e^{a_4\left(t_f - t_1^j\right)} \\[2mm] + \dfrac{A_{22}}{\tau_{ac}}e^{\frac{-\left(t_f - t_1^j\right)}{\tau_{ac}}} + \dfrac{A_{32}}{\tau_{de}}e^{\frac{-\left(t_f - t_1^j\right)}{\tau_{de}}} \end{array}\right).
\end{aligned}
\tag{3.91}
$$

Next, solve for t_1^{j+1}

$$
t_1^{j+1} = \frac{\left(\begin{array}{c}
\theta(t_f) - K_{11}e^{a_1 t_f} - K_{21}e^{a_2 t_f} - K_{31}e^{a_3 t_f} - K_{41}e^{a_4 t_f} - A_{11} - A_{21}e^{\frac{-t_f}{\tau_{ac}}} - A_{31}e^{\frac{-t_f}{\tau_{de}}} \\[2mm]
- K_{12}e^{a_1\left(t_f - t_1^j\right)}\left(1 - t_1^j a_1\right) - K_{22}e^{a_2\left(t_f - t_1^j\right)}\left(1 - a_2 t_1^j\right) - K_{32}e^{a_3\left(t_f - t_1^j\right)}\left(1 - t_1^j a_3\right) \\[2mm]
- K_{42}e^{a_4\left(t_f - t_1^j\right)}\left(1 - a_4 t_1^j\right) - A_{12} - A_{22}e^{\frac{-\left(t_f - t_1^j\right)}{\tau_{ac}}}\left(1 - \frac{t_1^j}{\tau_{ac}}\right) - A_{32}e^{\frac{-\left(t_f - t_1^j\right)}{\tau_{de}}}\left(1 - \frac{t_1^j}{\tau_{de}}\right)
\end{array}\right)}{\left(\begin{array}{c}
a_1 K_{12}e^{a_1\left(t_f - t_1^j\right)} + a_2 K_{22}e^{a_2\left(t_f - t_1^j\right)} + a_3 K_{32}e^{a_3\left(t_f - t_1^j\right)} + a_4 K_{42}e^{a_4\left(t_f - t_1^j\right)} \\[2mm]
+ \dfrac{A_{22}}{\tau_{ac}}e^{\frac{-\left(t_f - t_1^j\right)}{\tau_{ac}}} + \dfrac{A_{32}}{\tau_{de}}e^{\frac{-\left(t_f - t_1^j\right)}{\tau_{de}}}
\end{array}\right)}
\tag{3.92}
$$

where t_1^0 is the initial guess value of the switch-time and $(j + 1)^{th}$ iterates of t_1^{j+1} are calculated from the jth iterates of t_1^j.

The procedure of specifying the minimum-time controller begins by fixing t_f and using the linear approximation of Eq. (3.92) to solve for t_1^{j+1} and iterating until the desired degree of accuracy is achieved. Next, this procedure is repeated after decreasing t_f until the smallest t_f is found.

Enderle and Wolfe also examined the case of more than one switch time (Enderle and Wolfe, 1987). Their analysis demonstrated that the saccadic eye movement system is time-optimal with a single switch-time.

Presented in Fig. 3.40 are the optimal control results which illustrate the saccade magnitude

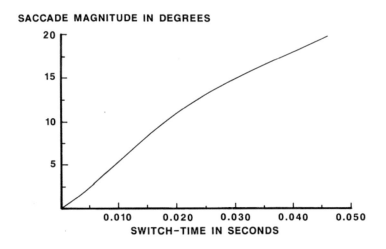

Figure 3.40: A diagram of the saccade magnitude and switch-time relationship for $F_p = 1.0$ N, $\tau_{ac} = 4$ ms, and $\tau_{de} = 5$ ms.

and switch-time relationship. The function is monotonically increasing with an inflection point at approximately 10°. For saccade magnitudes less than approximately 10°, the function is concave upward. For saccade magnitudes greater than approximately 10°, the function is concave downward. Time-optimal control results are presented in Fig. 3.41 for various values of τ_{ac} and $\tau_{de} = 5$ ms.

PEAK VELOCITY

The results of the data analysis on one of the three subjects discussed previously using the two-point central difference method are illustrated in Fig. 3.42. Velocity estimates are computed with a step size of 3 and a sampling interval of 1 ms. The time interval from the start of the saccade to the time at peak velocity t_{mv} showed marked variation within the 5, 10, and 15° target movements. The time at peak velocity should not be interpreted as a switching time. In fact, theoretical predictions in the previous section indicate that peak velocity occurs after the switching time for small eye movements and before the switching time for large eye movements. The range of variation on t_{mv} within each target movement is approximately constant and independent of the size of the target displacement.

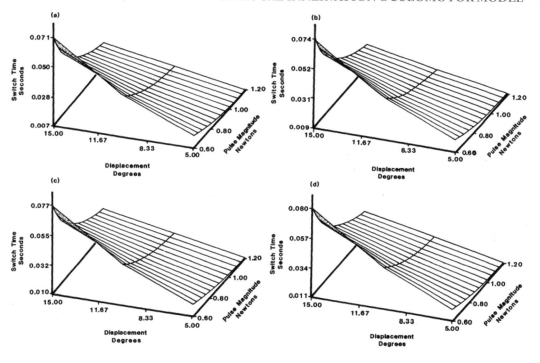

Figure 3.41: Diagram illustrating the effect of the activation time constant on the switch-time for first-order time-optimal neurological control signals. (a) τ_{ac} = 4 ms, (b) τ_{ac} = 7 ms, (c) τ_{ac} = 10 ms, (d) τ_{ac} = 13 ms, and τ_{de} = 5 ms.

Since the actual mechanical elements of the oculomotor system are not changing for saccades of the same size, the input to the oculomotor system must be responsible for the differences in saccade dynamics. As previously illustrated, the only parameter capable of changing the time to peak velocity is a variable activation time constant. Increasing the activation time constant predominantly increases the time to peak velocity while slightly reducing the peak velocity. These results are indicative of a random or variable activation time constant acting independently of saccade magnitude. Next, peak velocity varies greatly for all target displacements within the 5, 10, and 15° target movements. Peak velocity is influenced by the filter time constants, the magnitude of the agonist pulse, as well as the switching time. The most dominant factor affecting peak velocity, however, is the size of the agonist pulse magnitude. Increasing the size of the agonist pulse magnitude directly increases the peak velocity. This implies that the peak velocity variability apparent in the central difference results is primarily due to the magnitude of the agonist pulse, and that agonist pulse magnitude is a random variable, independent of the size of the target displacement. Note that the random behavior of these parameters is not a function of fatigue because the recording sessions were kept short to avoid fatigue.

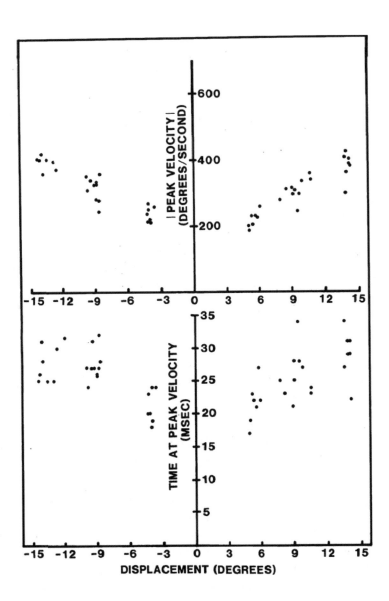

Figure 3.42: Two-point central difference estimates of peak velocity and time to peak velocity as a function of saccade magnitude. Velocity estimates are computed with a step size of 3 and a sampling interval of 1 ms.

SYSTEM IDENTIFICATION RESULTS

Displayed in Fig. 3.38 are the estimates of agonist pulse magnitude as a function of displacement for the three subjects tested. The estimated agonist pulse magnitude showed more variation within each target movement than between target movements. One pronounced feature evident from this graph is the apparent lack of a strong relationship between agonist pulse magnitude and saccade amplitude related by other investigators. In fact, pulse magnitude evidently does not depend on saccade size, consistent with a time-optimal control. Under a time optimal control, the magnitude of the agonist pulse should be a maximum regardless of the size of the saccade. Only the duration of the agonist pulse affects the size of the saccade. Figure 3.43 compares the average agonist pulse magnitude of

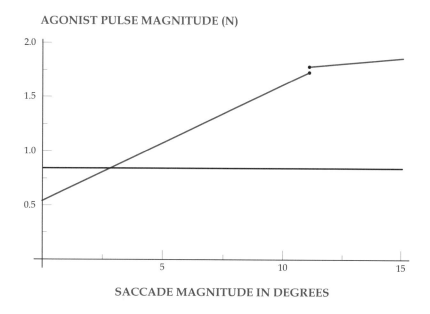

AGONIST PULSE MAGNITUDE (N)

SACCADE MAGNITUDE IN DEGREES

Figure 3.43: Diagram comparing the agonist pulse magnitude as a function of saccade magnitude as predicted by a first-order time-optimal control signal (blue line) and according to Bahill, A. (1981) (red line). Note that the first-order time-optimal agonist pulse magnitude is the average value from subject (a).

one of the three subjects tested to the agonist pulse magnitude according to Bahill et al. (1980). As indicated earlier, had the agonist pulse magnitude estimates displayed any dependence on saccade magnitude, as in Bahill's prediction, the controller would not have been time-optimal. Interestingly, Lehman and Stark use an agonist pulse magnitude that is a function of amplitude, a controller that they acknowledge as violating a bang-bang or optimal controller.

Clark and Stark first stated that the neuronal control strategy for human saccadic eye movements is time-optimal based on experimental data analysis (Clark and Stark, 1975). Using a nonlinear model with activation and deactivation time constants, they analyzed three different sets of agonist-antagonist controller inputs. Based on a curve fitting investigation matching model predictions to saccadic eye movement data, Clark and Stark concluded that the best results are obtained with a first-order pulse-step neuronal controller. Based on their optimal control investigation, however, they reported a second-order time optimal control signal. By reducing the order of the model from sixth to fourth-order, eliminating the activation and deactivation time constants, Clark and Stark's optimal control investigation yielded a first-order time optimal control signal, consistent with their experimental findings. Note that Clark and Stark did not solve for the switch-times in their optimal control analysis or comment on the pulse magnitude-saccade amplitude relationship. Lehman and Stark also investigated the neuronal control strategy for human saccadic eye movements (Lehman and Stark, 1979). Applying Pontryagin's minimum principle on a linear model, which includes the activation and deactivation time constants, failed to give robust results. After reducing the order of the model from sixth- to fourth-order, as Clark and Stark did by eliminating the activation and deactivation time constants, and applying Pontryagin's minimum principle, their analysis yielded a second-order time-optimal control signal. In simulating saccadic eye movements, Lehman and Stark, however, assumed that the agonist pulse magnitude is a function of saccade magnitude, a controller that is not time optimal since it violates Pontryagin's minimum principle.

Table 3.3: Saccadic Eye Movement Time-Optimal Controller Results with the Activation and Deactivation Time Constants Included in the Analysis.

Investigator	Model	Time-Optimal Controller
Clark and Stark	Sixth-Order Nonlinear	Second-Order
Lehman and Stark	Sixth-Order Linear	Unable to Specify
Enderle and Wolfe	Sixth-Order Linear	First-order

Table 3.4: Saccadic Eye Movement Time-Optimal Controller Results without the Activation and Deactivation Time Constants Included in the Analysis.

Investigator	Model	Time-Optimal Controller
Clark and Stark	Fourth-Order Nonlinear	First-Order
Lehman and Stark	Fourth-Order Linear	Second-Order

In omitting the activation and deactivation time constants in their optimal control investigation, but not in their simulations, both Clark and Stark (1975) and Lehman and Stark (1979) implicitly assume that the values of the time constants are zero. However, there is abundant physiological evidence for including activation and deactivation time constants in models of saccadic eye movements (Robinson, D., 1981; Bahill, A., 1981). Additionally, sensitivity analyses indicate that both of these time constants are important but not dominant, control parameters (Clark and Stark, 1975; Hsu et al., 1976). It therefore seems important that these time constants be included in the neurological control investigation of saccadic eye movements.

Tables 3.3 and 3.4 summarize the theoretical time-optimal saccadic eye movement control results reported here in this section. Each of the investigators uses a fourth-order oculomotor plant in their analysis. The difference between the tables involves including the activation and deactivation time constants in the optimal control analysis (which increases the order of the system from fourth- to a sixth-order) or not including the time constants. Furthermore, it should be noted that Lehman and Stark's model is linearized from Clark and Stark's model.

In comparing the table listings, it is apparent that the optimal controller for the nonlinear oculomotor system is different from the optimal controller for the linear oculomotor model. The differences in the linear and nonlinear time-optimal controllers are probably not attributable to the linearization because of the high degree of accuracy exhibited by the linear model. Such differences might be attributed to the assumptions regarding the costate variables p_2 and p_3 by Clark and Stark. Lehman and Stark report that the costate variables are extremely sensitive to the initial conditions, which can only be roughly estimated. Further, Lehman and Stark state that varying the costate variable initial conditions results in different order controllers. Thus, the theoretical time-optimal controller specified by Clark and Stark is probably in error due to the lack of information about the costate variables initial conditions and since it did not agree with their observed first-order time-optimal control simulation solution. Only the theoretical first-order time-optimal control results presented here includes the activation and deactivation time constants in the oculomotor system model and agree with the experimental results of Clark and Stark (1975), and Enderle and Wolfe (1988).

Example 3.3

Using the oculomotor plant model described with Eq. (3.49) and the following parameters, simulate a 20° saccade.

$K_{se} = 125 \text{ Nm}^{-1}$

$K_{lt} = 32 \text{ Nm}^{-1}$

$K = 66.4 \text{ Nm}^{-1}$

$B = 3.1 \text{ Nsm}^{-1}$

$J = 2.2 \times 10^{-3} \text{ Ns}^2\text{m}^{-1}$

$B_{ag} = 3.4 \text{ Nsm}^{-1}$

$B_{ant} = 1.2 \text{ Nsm}^{-1}$

$\tau_{ac} = 0.009 \text{ s}$

$\tau_{de} = 0.0054 \text{ s}$

$\delta = 5.80288 \times 10^5$

$F_p = 1.3 \text{ N}$

$t_1 = 31$ ms

Latent Period $= 150$ ms.

Plot the neural inputs, agonist and antagonist active-state tension, position, velocity and acceleration vs. time.

Solution.

Using the given parameter values, the following m-file provides the parameter values for the Simulink program.

```
Fp=1.3
t1=.031
theta=20
theta0=0
TDE=.0054
TAC=.009
tdeinv=1/TDE
tacinv=1/TAC
KSE=125
KLT=32
K=66.4
B=3.1
J=2.2 * 10^{}-3
BAG=3.4
BANT=1.2
DELTA=5.80261*10^5
KST=KLT+KSE
C0=((K*KST^2)+(2*KSE*KST*KLT))/(J*BANT*BAG)
C1=((B*KST^2)+(BAG+BANT)*((K*KST)+(2*KSE*KST)-KSE^2)))/(J*BANT*BAG)
C2=((J*KST^2)+((B*KST)*(BAG+BANT))+((BANT*BAG)*(K+(2*KSE))))/(J*BANT*BAG)
C3=(((J*KST)*(BAG+BANT))+(B*BANT*BAG))/(J*BANT*BAG)
if theta0 $<$ 14.23
   Fag0=0.14+0.0185*theta0
else
   Fag0=0.0283*theta0
end
if theta0 < 14.23
   Fant0=0.14-0.0098*theta0
else
   Fant0=0
end
```

```
if theta < 14.23
    Fagss=0.14+0.0185*theta
else
    Fagss=0.0283*theta
end
if theta < 14.23
    Fantss=0.14-0.0098*theta
else
    Fantss=0
end
latent=.15
sstart=latent+t1
agstep=Fp-Fagss
```

Shown in Fig. 3.44 is the Simulink program. The main program is shown in Fig. 3.44 (A) based on Eq. (3.49). The input to the system is shown in Fig. 3.44 (B), the agonist and antagonist input is shown in Fig. 3.44 (C) and (D). In Fig. 3.45 are plots of position, velocity, acceleration, agonist neural input and active state tension, and antagonist neural input and active state tension.

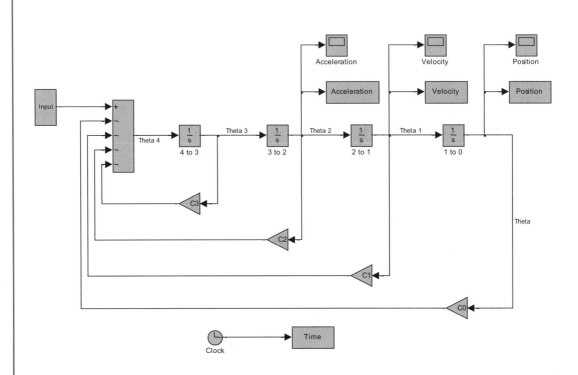

Figure 3.44: Simulink program for Example 3.3 (A) (Main Simulink Program).

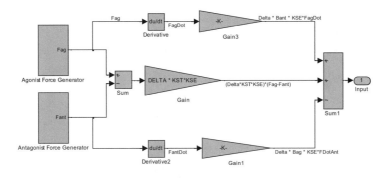

Figure 3.44: (continued). Simulink program for Example 3.3 (B). (Input).

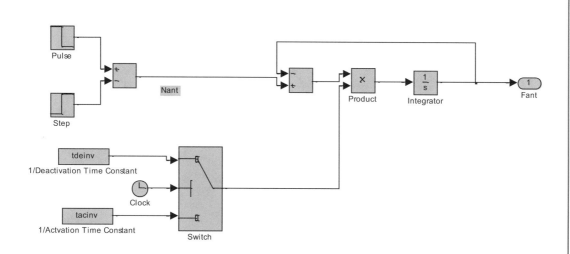

Figure 3.44: (continued). Simulink program for Example 3.3 (C) (Agonist Input).

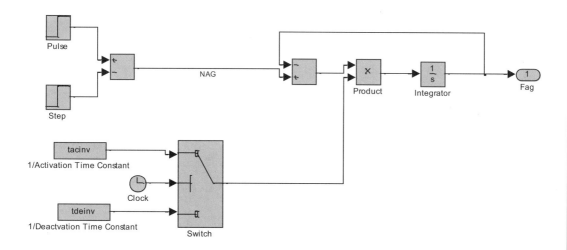

Figure 3.44: (continued). Simulink program for Example 3.3 (D). (Antagonist Input).

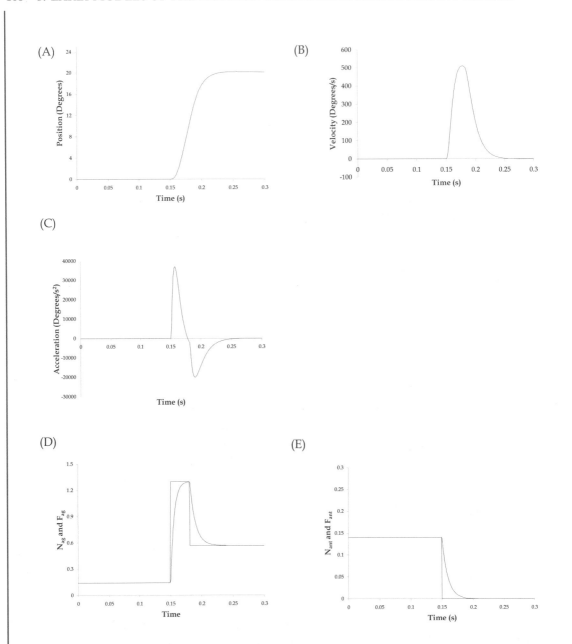

Figure 3.45: Plots of position, velocity, acceleration, agonist neural input and active state tension, and antagonist neural input and active state tension for Example 3.2.

CHAPTER 4

Velocity and Acceleration Estimation

4.1 INTRODUCTION

Estimates of accurate saccade velocity and acceleration are essential in the study of the oculomotor system. Saccade velocities are used in clinical studies and studies of the frequency characteristics of saccade neural networks and the oculomotor plant. Most studies use the main sequence diagram that describes the relationship between peak velocity versus saccade amplitude. The accuracy of velocity and acceleration estimates from three algorithms is presented in this section. The saccade signal is composed of low frequency components that are contaminated by biological noise as well as noise caused by the use of measuring devices, quantization, and analog-to-digital conversion. The choice of a derivative algorithm depends on simplicity, desired accuracy and the frequency characteristics.

In verifying a model, it is often helpful to examine global characteristics. For instance, when the Westheimer model was introduced, we examined time to peak velocity and peak velocity as functions of the size of the saccade. Another important characteristic of this system is the latent period, that is, the time it takes from the movement of the target to the start of the eye movement. Saccade peak velocity-saccade magnitude and duration characteristics are usually referred to as the main sequence characteristics.

4.2 TWO-POINT CENTRAL DIFFERENCE METHOD

One method for computing derivatives is the central difference method. For estimators of two data points, it is the best method as it introduces no phase shifts. To compute the first derivative of a sequence of data points according to the central difference method, we calculate

$$\dot{y}(kT) = \frac{y((k+n)T) - y((k-n)T)}{2nT} \tag{4.1}$$

where T is the sampling interval (s), nT is the step size and k is the discrete time variable. Note that the experimenter selects T and n according to the system studied and the degree of accuracy desired.

4.2.1 FREQUENCY CHARACTERISTICS

The accuracy of the derivative approximation is best seen in the frequency domain. Note that the central difference algorithm differentiates and filters — a very desirable feature when analyzing data.

Consider a general function of time $x(t)$ and its derivative $\dot{x}(t)$. In the frequency domain, the relationship between the two is:

$$\dot{X}(j\omega) = j\omega X(j\omega) .$$

Now a two-point central difference estimate of the derivative is

$$\dot{y}(kT) = \frac{y((k+1)T) - y((k-1)T)}{2T}$$

and the z-transform is

$$\dot{Y}(z) = Y(z)\frac{\left(z - z^{-1}\right)}{2T} .$$

We can return to the frequency domain by letting $z = e^{j\omega T} = \cos(\omega t) + j\sin(\omega t)$,

$$\dot{Y}(\omega T) = Y(\omega T)\left(\frac{e^{j\omega t} - e^{-j\omega t}}{2T}\right)$$

yielding

$$\dot{Y}(\omega T) = Y(\omega T)\frac{j\sin\omega T}{T} .$$

Now ideally

$$\frac{\dot{X}(j\omega)}{X(j\omega)} = j\omega \quad \text{or} \quad \left|\frac{\dot{X}(\omega)}{X(\omega)}\right| = \omega$$

and experimentally

$$\frac{\dot{Y}(\omega T)}{Y(\omega T)} = \frac{j\sin\omega T}{T} \quad \text{or} \quad \left|\frac{\dot{Y}(\omega T)}{Y(\omega T)}\right| = \frac{\sin\omega T}{T} .$$

The graph in Fig. 4.1 is of the ABS (Gain) for the true derivative and the estimated central difference derivative vs. frequency with $T = 0.001$ s. The bandwidth (3 db point) is approximately 221 Hz, (found from a Bode plot). Note that we typically low-pass filter (analog filter) frequencies above the highest frequency content of our signal to reduce *aliasing*. Based on the work of Bahill and McDonald (1983b), the relationship between the bandwidth and the amount of spread between points for the central difference method is expressed as:

$$\text{bandwidth (Hz)} = \frac{0.443 \times \text{sampling rate (Hz)}}{\text{spread}}$$

that is, ± 3 points (a spread of 6) yields a bandwidth of 74 Hz.

The maximum frequency content for saccade velocity is estimated at 74 Hz, and 45 Hz for saccade acceleration. Stated simply, the optimal estimate is found by having the smallest bandwidth for the central difference algorithm without losing any of the signal. Thus, for saccades, we sample with $T = 0.001$ s, and can calculate a velocity estimate as

$$\dot{y}(kT) = \frac{y((k+3)T) - y((k-3)T)}{6T}$$

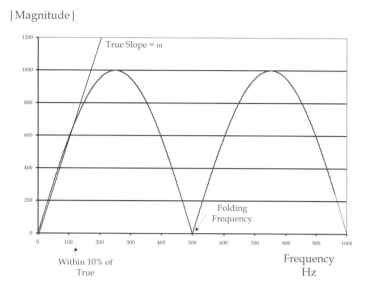

Figure 4.1: Plot of gain for the true derivative and the estimated central difference derivative vs. frequency.

which has a bandwidth of 74 Hz, and an acceleration estimate as

$$\ddot{y}(kT) = \frac{\dot{y}((k+4)T - \dot{y}(k-4)T)}{8T}$$

which has a bandwidth of 55 Hz.

4.3   BAND-LIMITED DIFFERENTIATION FILTER

The choice of a derivative filter depends on simplicity, accuracy, and frequency characteristics. With the computer power available to us today, relying on simple two-point filters because of computational needs is no longer a major concern. The choice of the derivative filter today should be dependent on the desired degree of accuracy and the frequency characteristics.

One excellent derivative filter is called a band-limited derivative (BLD) filter, which is of the form:

$$y(kT) = \sum_{n=-N}^{N} C_n x((k-n)T) \tag{4.2}$$

where x_n is the input data sequence, y_n is the derivative sequence, N is the number of input values used on each side of the current input value, and C_k is the filter coefficients. The filter is a form of nonrecursive filter as it does not use past values of the filtered output, but it merely depends on the input. This filter can be used to calculate time derivatives with a phase response that is linear.

Since y_n is calculated after N future input values are sampled and stored, a linear phase shift results between the input and the filtered data with this filter. If a real time filter is desired, we have no knowledge of future data $(x_{n+1}$ to $x_{n+N})$ and, therefore, this filter is not appropriate.

The transfer function of a band-limited differentiating filter is shown in Fig. 4.2. Notice that

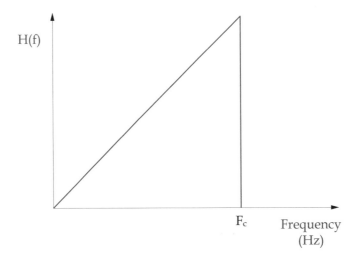

Figure 4.2: The transfer function of a BLD filter.

function has the shape of an ideal differentiator until the frequency F_c; afterwards, it acts like a low-pass filter that removes any signal content above F_c. The function is a ramp function with slope of 2π up to the cutoff frequency, F_c, above which, it is zero.

To create a BLD filter with the characteristics of the above transfer function, the coefficients are obtained from a Fourier-series expansion of the ideal transfer function in Fig. 4.2 as follows

$$C_k = 2 \int_0^{F_c} 2\pi f \, \sin(2\pi k f) df$$

which reduces to

$$C_k = \frac{1}{\pi} \left(\frac{\sin(2\pi k F_c)}{k^2} - \frac{2\pi F_c \cos(2\pi k F_c)}{k} \right).$$

The filter coefficients from the previous equation are calculated in the range of $-N \leq k \leq N$. Note that because the filter coefficients are antisymmetric, $C_k = C_{-k}$ and $C_0 = 0$. The Fourier-series expansion also assumes a sampling interval of unity. Therefore, a factor of $\frac{1}{T}$ is used to scale the coefficients.

$$\dot{y}(kT) = \frac{1}{T} \sum_{n=1}^{N} C_n \left(y\left((k+n)T\right) - y\left((k-n)T\right) \right).$$

The sampling rate for saccades is typically $T = 0.001$ s. Using power spectral analysis, the cutoff frequency can be determined by setting F_c such that 99% of the energy falls within it. A value of approximately $F_c = 30$ Hz is appropriate for most saccades. A good value for N is 45 points.

Truncating the filter at 45 points causes an undesirable characteristic know as the Gibbs phenomenon to appear in the filtered response. The Gibbs response is a ripple. To control the ripple, a Kaiser window is used to reduce the effect of the Gibbs phenomenon at the cost of increasing the transition zone. The Kaiser window is a flexible window that allows for the adjustment via parameter α to produce the ripple and transition zone desired. The Kaiser window weights are given by:

$$w_k = \frac{I_0\left(\alpha\sqrt{1 - \left(\frac{n}{N}\right)^2}\right)}{I_0\left(\alpha\right)}, \quad |n| \le N$$

and

$$I_0\left(x\right) = 1 + \sum_{n=1}^{\infty}\left(\frac{\left(\frac{x}{2}\right)^n}{n!}\right)^2$$

where α is the parameter used to reduce the Gibbs Phenomenon. For saccades, a value of $\alpha = 5.4414$ works well.

Combining the Kaiser window with the other filter gives the following equation for estimation of velocity of a saccade:

$$\dot{y}\left(kT\right) = \frac{1}{T}\sum_{n=1}^{N}w_n C_n\left(y\left(\left(k + n\right)T\right) - y\left(\left(k - n\right)T\right)\right).$$

The following is a FORTRAN program for calculating the saccade velocity with the BLD filter. Note that the coefficients used for $\omega_n C_n$ are given in the data statement for vcoef.

```
 real*4 time(500),position(500),velocity(500),vcoef(45),vdata(500)
+ ,accel(500),adata(500)
  data vcoef/.7072319E+00,.1394343E+01,.2042004E+01,
+.2632429E+01,.3150056E+01,.3582120E+01,.3919101E+01,
+.4155012E+01,.4287543E+01,.4318018E+01,.4251214E+01,
+.4095032E+01,.3860031E+01,.3558882E+01,.3205744E+01,
+.2815608E+01,.2403643E+01,.1984566E+01,.1572080E+01,
+.1178384E+01,.8137948E+00,.4864816E+00,.2023151E+00,
+-.3516010E-01,-.2246426E+00,-.3668937E+00,-.4644756E+00,
+-.5214275E+00,-.5429054E+00,-.5348019E+00,-.5033759E+00,
+-.4549027E+00,-.3953669E+00,-.3302089E+00,-.2641316E+00,
+-.2009733E+00,-.1436451E+00,-.9412885E-01,-.5352848E-01,
+-.2216477E-01,.2980921E-03,.1470597E-01,.2226023E-01,
+.2436629E-01,.2249453E-01/
```

```
 open(10,file=
+'Model Simulation Output Time and Position-Large Noise.txt',
+status='old',form='formatted')
     do 225 i=1,400
            read(10,226)time(i),position(i)
225          continue
226 format(F10.4,F11.4)
    tsampl=.001
  ncoef=45
    maxit=350
  ibegin=50
  npt=400
     open(12,file='bld.txt',form='formatted')
c Compute velocity estimates
c
c     BAND LIMITED DIFFERENTIATION
        do 270 i=ibegin,maxit
           k=i-ibegin+1
           vdata(k)=0.
         do 260 j=1,ncoef
              vdata(k)=vdata(k)+vcoef(j)*(position(i+j)-position(i-j))
260        continue
        velocity(i)=vdata(k)
270        continue
  do 370 i=100,301
       k=i-ibegin+1
       adata(k)=0.
       do 360 j=1,ncoef
            adata(k)=adata(k)+vcoef(j)*(velocity(i+j)-velocity(i-j))
360      continue
      accel(i)=adata(k)
370    continue
      do 470 i=101,301
   write(12,227)time(i),position(i),velocity(i),accel(i)
470  continue
227    format(4F15.4)
      stop
      end
```

4.4   MEDIAN DIFFERENTIATION FILTER

The BLD filter has the advantage of better bandwidth control and superior pass band and stop-band performance than that of the two-point central difference method. The disadvantage of the BLD is the long impulse response. This is the classic linear filter trade-off between the frequency domain and the time domain. The impulses typically produce a ringing in the filter output. This ringing is simply the impulse response of the filter evoked by the impulse noise. It is an inherent characteristic of all linear finite impulse response and infinite impulse response filters. The ideal differentiating filter has a short impulse response of the two-point central difference method and the stop-band performance of the BLD.

Estimating velocity using the BLD filter greatly magnifies low level impulse noise that occasionally appears in the output of eye movement measurements. A direct way to avoid this problem is to use a differentiating filter that has no impulse response function, which requires a nonlinear filter. This is because a linear filter with no impulse response will have no response for any input. A median filter, a class of order statistic filters, has no impulse response and no transfer function. Strictly speaking, the terms *frequency response* and *bandwidth* do not apply to the median filter.

Order statistic filters are a class of nonlinear digital filters that operate on a sliding window of input data samples. Usually the window is of odd length; i.e., $L = 2N + 1$. The samples in the window are rank-ordered and the ordered samples (ordered statistics) are linearly weighted. This linear combination of order statistics constitutes the filter output. For example, the standard median filter is an order statistic filter in which the center sample of the ordered array is given a weight of *1* and all other samples a weight *0*. Other order statistic filters are implemented by choosing different weighting coefficients. The key element in all order statistic filters is the rank ordering operation, which is a data dependent, nonlinear process. Order statistic filters are nonlinear since the principle of superposition does not apply.

The median filter may be introduced into a differentiating filter in two ways. The nonlinear operation (median) may be applied first followed by the linear operation (differentiation). The reverse also works. In this case, the differentiation is performed before the median operation. The latter approach is discussed in this section. This approach requires several independent derivative estimates to be calculated so that an unbiased median is found. Here the two point central difference method is used to calculate the derivative. The median derivative filter is a classic order-statistical filter consisting of a bank of linear filters (two-point differences) followed by the median operation and given by

$$\overset{\bullet}{y}(kT) = \frac{1}{2nT} MEDIAN \begin{cases} y((k+6)T) - y(kT), \ y((k+5)T) - y((k-1)T), \\ y((k+4)T) - y((k-2)T), \ y((k+3)T) - y((k-3)T), \\ y((k+2)T) - y((k-4)T), \ y((k+1)T) - y((k-5)T), \\ y(kT) - y((k-6)T). \end{cases}$$

While the median filter does not have a *bandwidth*, a bandwidth can be estimated based on the linear component of the filter (two-point central difference estimate). The nonlinear component of the median filter has a *low-pass filter*-like property. Therefore, the median filter can be considered a cascade combination of a low-pass filter and a differentiator. For design purposes, the bandwidth of the median filter is essentially the two-point central difference estimate, and the impulse rejection property is related to the number of differences used in the median operation. The more differences used, the greater the impulse rejection capability.

The following is a FORTRAN program for calculating the saccade velocity with the median filter.

```
USE MSIMSL
    real*4 time(500),position(500),velocity(500),ra(7),rb(7),
    + median,accel(500)
    open(10,file=
    +'Model Simulation Output Time and Position-Large Noise.txt',
  +status='old',form='formatted')
      do 225 i=1,400
            read(10,226)time(i),position(i)
225         continue
226    format(F10.4,F11.4)
    tsampl=.001
c nsample is the number of eye movements to be plotted
    maxit=350
    ibegin=50
    npt=400
      open(12,file='median.txt',form='formatted')
c    Median Derivative Method
c
870    do 809 i=ibegin,maxit
      ra(1)=position(i+6)-position(i)
      ra(2)=position(i+5)-position(i-1)
      ra(3)=position(i+4)-position(i-2)
      ra(4)=position(i+3)-position(i-3)
      ra(5)=position(i+2)-position(i-4)
      ra(6)=position(i+1)-position(i-5)
      ra(7)=position(i)-position(i-6)
      call SVRGN(7,ra,rb)
      median=rb(4)
      velocity(i)=median/(6*.001)
809    continue
      do 810 i=1,301
      ra(1)=velocity(i+6)-velocity(i)
```

```
        ra(2)=velocity(i+5)-velocity(i-1)
        ra(3)=velocity(i+4)-velocity(i-2)
        ra(4)=velocity(i+3)-velocity(i-3)
        ra(5)=velocity(i+2)-velocity(i-4)
        ra(6)=velocity(i+1)-velocity(i-5)
        ra(7)=velocity(i)-velocity(i-6)
        call SVRGN(7,ra,rb)
        median=rb(4)
        accel(i)=median/(6*.001)
810     continue
        do 270 i=101,301
    write(12,227)time(i),position(i),velocity(i),accel(i)
270     continue
227     format(4F15.4)
        stop
        end
```

4.5   SUMMARY OF DIFFERENTIATION RESULTS

Shown in the following figures are the simulation results and filter estimates for velocity and acceleration. Noise was added to the simulated eye position to create realistic eye movement data.

Since the two-point central difference method (TPCD) is the differentiation method of choice for most researchers in saccadic eye movement analysis, results obtained using two other techniques are compared to those results. In general, little filtering of the noise is carried using the TPCD filter since the transfer function is similar to a rectified sine wave, and noise is greatly amplified as the order of the derivative increases. The BLD filter provides best low-pass filtering above the cutoff frequency since it is based on a frequency domain triangle waveform. However, while the BLD filter greatly eliminates the noise, it does so at the expense of inaccurately tracking the true waveform. The BLD is also the most computationally expensive as compared to the other two techniques.

As illustrated in Fig. 4.3, the median filter velocity estimate provides the most accurate representation of the simulated velocity signal, especially during the saccade. The median filter provides the best estimate of peak velocity of all of the filters, with the TPCD overestimating and the BLD underestimating. Similarly, the median filter acceleration estimate is the most accurate of the three algorithms as shown again in Fig. 4.4. The BLD severely underestimates the peak acceleration. The TPCD provides a poor estimate during fixation, and a noisy estimate during the saccade. The median filter provides the best fixation estimate of acceleration and the closest match to the acceleration during the saccade. However, it falls short of providing a good estimate of peak acceleration. Similar results are observed for 5°, 15°, and 20° saccades.

The median derivative filtering method is demonstrated to provide many advantages over the two-point central difference and the band-limited derivative method in computing velocity and

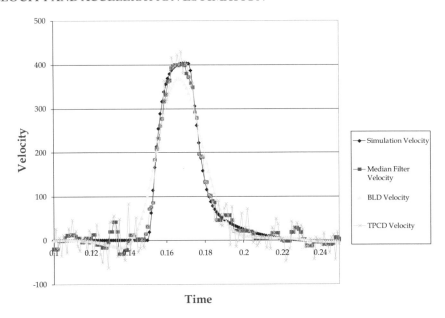

Figure 4.3: Saccade velocity calculated from the oculomotor model and three estimates.

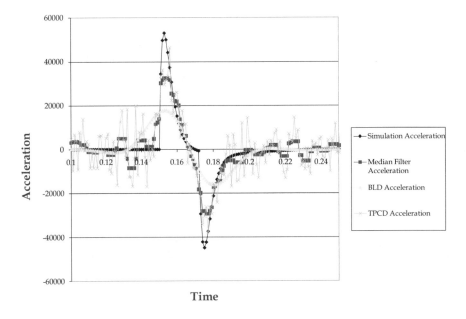

Figure 4.4: Saccade acceleration calculated from the oculomotor model and three estimates.

acceleration estimates. The two-point central difference method is a two-tap finite impulse response filter, and thus suffers from an impulse response. The BLD filter has the advantage of superior pass and stop band performance, but it suffers from a long impulse response (transient response), which is the classic trade-off between frequency and time domain performance. The median derivative filter is a nonlinear filter, which has no impulse response and a bandwidth defined by the low-pass differentiator (Engelken et al., 1996, 1993, 1991, 1990). Improved impulse rejection is dependent on the number of differences used in the median operation. This means that the more differences, the better the impulse rejection capability.

The main sequence diagram is the relationship between peak velocity and saccade amplitude. As shown, the median filter provides the best estimate of peak velocity among the three filters. Clinically, saccades are defined as too slow or fast if they fall outside bounds defined from the main sequence diagram. Saccades that are too fast are typified by patients with myasthenia gravis. Slow saccades usually occur in patients with ocular muscle or ocular motor nerve paresis, or central neurological disorders. Saccade velocity is lower in drowsy, inattentive, drugged, intoxicated, or aged patients.

CHAPTER 5

1995 Linear Homeomorphic Saccadic Eye Movement Model

5.1 INTRODUCTION

In Chapter 3, we presented a linear model of the oculomotor plant developed by linearizing the force-velocity curve. We then derived a linear differential equation to describe the system. Here we re-examine the static and dynamic properties of muscle in the development of a linear model of oculomotor muscle. With the updated linear model of oculomotor muscle, the model of the oculomotor system will also be updated.

5.2 LINEAR MUSCLE MODEL

The updated linear model for oculomotor muscle is shown in Fig. 5.1. Each of the elements in the model is linear and supported with physiological evidence. The muscle is modeled as a parallel combination of viscosity B_2 and series elasticity K_{se}, connected to the parallel combination of active state tension generator F, viscosity element B_1, and length tension elastic element K_{lt}. Variables x_1 and x_2 describe the displacement from the equilibrium for the stiffness elements in the muscle model. The only structural difference between this model and the previous oculomotor muscle model is the addition of viscous element B_2 and the removal of passive elasticity K_{pe}. As will be described, the viscous element B_2 is vitally important to describe the nonlinear force-velocity characteristics of the muscle, and the elastic element K_{pe} is unnecessary.

The need for two elastic elements in the linear oculomotor muscle model is supported through physiological evidence. As described previously, the use and value of the series elasticity K_{se} was determined from the isotonic-isometric quick release experiment by Collins. Length-tension elasticity K_{lt} was estimated in a slightly different fashion than before from the slope of the length-tension curve. Support for the two linear viscous elements is based on the isotonic experiment and estimated from simulation results presented in this chapter.

5.2.1 LENGTH-TENSION CURVE

The basis for assuming nonlinear elasticity is the nonlinear length-tension relation for excited and unexcited muscle for tensions below 10 g as shown in Fig. 3.8. Using a miniature "C" gauge force transducer, Collins in 1975 (Collins, C., 1975) measured muscle tension *in vivo* at the muscle tendon during unrestrained human eye movements. Data shown in Fig. 3.8 were recorded from the rectus

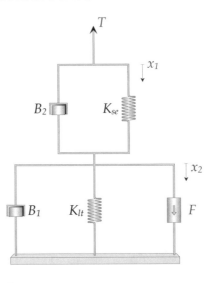

Figure 5.1: Diagram illustrates an updated linear muscle model consisting of an active state tension generator F in parallel with a length-tension elastic element K_{lt} and viscous element B_1, connected to a series elastic element K_{se} in parallel with a viscous element B_2. Upon stimulation of the active state tension generator F, a tension T is exerted by the muscle.

muscle of the left eye by measuring the isometric tensions at different muscle lengths, ranging from eye positions of $-45°$ to $45°$, and different levels of innervation, established by directing the subject to look at the corresponding targets with the unhampered right eye from $-45°$ temporal (T) to $45°$ nasal (N). The change in eye position during this experiment corresponds to a change in muscle length of approximately 18 mm. Collins described the length-tension curves as "straight, parallel lines above about 10 g. Below the 10 g level, the oculorotary muscles begin to go slack." He also reported that the normal range of tensions for the rectus muscle during all eye movements *never* falls below 10 g into the slack region when the *in vivo* force transducer is used.

In developing a muscle model for use in the oculomotor system, it is imperative that the model accurately exhibits the static characteristics of rectus eye muscle within the normal range of operation. Thus, any oculomotor muscle model must have length-tension characteristics consisting of straight, parallel lines above 10 g tension. Since oculomotor muscles do not operate below 10 g, it is *unimportant* that the linear behavior of the model does not match this nonlinear portion in the length-tension curves observed in the data as was done in the development of the muscle model earlier. As demonstrated in this section, by concentrating on the operational region of the oculomotor muscles, accurate length-tension curves are obtained from the muscle model using just series elastic and length tension elastic elements, even when active state tension is zero. Thus, there is no need to include a passive elastic element in the muscle model as previously required.

Since the rectus eye muscle is not in equilibrium at primary position (looking straight ahead, $0°$) within the oculomotor system, it is necessary to define and account for the equilibrium position of the muscle. Equilibrium denotes the unstretched length of the muscle when the tension is zero, with zero input. It is assumed that the active state tension is zero on the $45°$ T length-tension curve. Typically, the equilibrium position for rectus eye muscle is found from within the slack region, where the $45°$ T length-tension curve intersects the horizontal axis. Note that this intersection point was not shown in the data collected by Collins (Fig. 3.8) but is reported to be approximately $15°$ (3 mm short of primary position), a value that is typical of those reported in the literature.

Since the muscle does not operate in the slack region during normal eye movements, using an equilibrium point calculated from the operational region of the muscle provides a much more realistic estimate for the muscle. Here, the equilibrium point is defined according to the straight-line approximation to the $45°$ T length-tension curve above the slack region. The value at the intersection of the straight-line approximation with the horizontal axis gives an equilibrium point of $-19.3°$. By use of the equilibrium point at $-19.3°$, there is no need to include an additional elastic element K_{pe} to account for the passive elasticity associated with unstimulated muscle as others have done.

The tension exerted by the linear muscle model shown in Fig. 5.1 is given by

$$T = \frac{K_{se}}{K_{se} + K_{lt}} F - \frac{K_{se} K_{lt}}{K_{se} + K_{lt}} x_1 . \tag{5.1}$$

With the slope of the length-tension curve equal to $0.8 \frac{g}{°} = 40.86 \frac{N}{m}$ in the operating region of the muscle (non slack region), $K_{se} = 2.5 \frac{g}{°} = 125 \frac{N}{m}$, and Eq. (5.1) has a slope of

$$\frac{K_{se} K_{lt}}{K_{se} + K_{lt}} . \tag{5.2}$$

K_{lt} is evaluated as $1.2 \frac{g}{°} = 60.7 \frac{N}{m}$.

To estimate the static active state tension for fixation at the locations detailed in Figure 1 of Collins (Collins, C., 1975), use the techniques by Enderle and coworkers (1991) by taking Eq. (5.1) to solve for steady-state active state tensions for each innervation level straight line approximation, yielding for $\theta > 0°$ (N direction)

$$F = 0.4 + 0.0175\theta \ N \quad \text{for } \theta > \ 0° \ (\text{N direction}) \tag{5.3}$$

and

$$F = 0.4 + 0.012\theta \ N \quad \text{for } \theta \ \leq \ 0° \ (\text{T direction}) \tag{5.4}$$

where θ is the angle that the eyeball is deviated from the primary position measured in degrees, and $\theta = 5208.7 \times (x_1 - 3.705)$. Note that $5208.7 = \frac{180}{\pi r}$, where r equals the radius of the eyeball with a value of 11 mm.

Figure 5.2 displays a family of static length-tension curves obtained using Eqs. (5.1)–(5.4), which depicts the length-tension experiment. No attempt is made to describe the activity within the

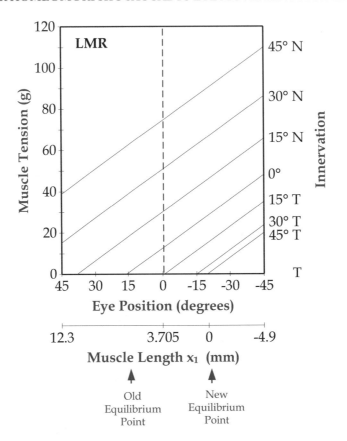

Figure 5.2: Length-tension relationship generated using Eqs. (5.1)–(5.4) derived from the linear muscle model and inputs: $F = 130$ g for 45° N, $F = 94.3$ g for 30° N, $F = 64.9$ g for 15° N, $F = 40.8$ g for 0°, $F = 21.7$ g for 15° T, $F = 5.1$ g for 30° T, and $F = 0$ g for 45° T. These lines were parameterized to match Fig. 3.8.

slack region since the rectus eye muscle does not normally operate in that region. The length-tension relationship shown in Fig. 5.2 is in excellent agreement with the data shown in Fig. 3.8 within the operating region of the muscle.

5.2.2   FORCE-VELOCITY RELATIONSHIP

The original basis for assuming nonlinear muscle viscosity is that the expected linear relation between external load and maximum velocity was not observed in early experiments by Fenn and Marsh. As Fenn and Marsh reported, "If the muscle is represented accurately by a viscous elastic system this force-velocity curve should have been linear, the loss of force being always proportional to the velocity. The slope of the curve would then represent the coefficient of viscosity." Essentially the same experiment was repeated for rectus eye muscle by Close and Luff (1974) with similar results.

The classic force-velocity experiment was performed to test the viscoelastic model for muscle as described previously in Section 3.4.4. Under these conditions, it was first assumed that the inertial force exerted by the load during isotonic shortening could be ignored. The second assumption was that if mass was not reduced enough by the lever ratio (enough to be ignored), then taking measurements at maximum velocity provided a measurement at a time when acceleration is zero, and, therefore, inertial force equals zero. If these two assumptions are valid, then the experiment would provide data free of the effect of inertial force as the gravity force is varied. Both assumptions are incorrect. The first assumption is wrong since the inertial force is never minimal (minimal would be zero) and therefore has to be taken into account. The second assumption is wrong since, given an inertial mass not equal to zero, then maximum velocity depends on the forces that act prior to the time of maximum velocity. The force-velocity relationship is carefully re-examined with the inertial force included in the analysis in this section.

The dynamic characteristics for the linear muscle model are described with a force-velocity curve calculated via the lever system presented in Fig. 3.14 and according to the isotonic experiment. For the rigid lever, the displacements x_1 and x_3 are directly proportional to the angle θ_1 and to each other, such that

$$\theta_1 = \frac{x_1}{d_1} = \frac{x_3}{d_3} . \tag{5.5}$$

The equation describing the torques acting on the lever is given by

$$Mgd_3 + Md_3^2\ddot{\theta}_1 = d_1 K_{se} (x_2 - x_1) + d_1 B_2 (\dot{x}_2 - \dot{x}_1) . \tag{5.6}$$

The equation describing the forces at node 2, inside the muscle, is given by

$$F = K_{lt}x_2 + B_1\dot{x}_2 + B_2 (\dot{x}_2 - \dot{x}_1) + K_{se} (x_2 - x_1) . \tag{5.7}$$

Equation (5.6) is rewritten by removing θ_l using Eq. (5.5), hence

$$Mg\frac{d_3}{d_1} + M\left(\frac{d_3}{d_1}\right)^2 \ddot{x}_1 = K_{se} (x_2 - x_1) + B_2 (\dot{x}_2 - \dot{x}_1) . \tag{5.8}$$

Ideally, to calculate the force-velocity curve for the lever system, $x_1(t)$ is found first. Then $\dot{x}_1(t)$ and $\ddot{x}_1(t)$ are found from $x_1(t)$. Finally, the velocity is found from $V_{max} = \dot{x}_1(T)$, where time T is the time it takes for the muscle to shorten to the stop, according to the experimental conditions of Close and Luff (1974). While this velocity may not be maximum velocity for all data points, the symbol V_{max} is used to denote the velocities in the force-velocity curve for ease in presentation. Note that this definition of velocity differs from the Fenn and Marsh, 1935 definition of velocity. Fenn and Marsh denoted maximum velocity as $V_{max} = \dot{x}_1(T)$, where time T is found when $\ddot{x}_1(T) = 0$.

It should be noted that this is a third-order system and the solution for $x_1(t)$ is not trivial and involves an exponential approximation (for an example of an exponential approximation solution for

V_{max} from a fourth-order model, see the paper by Enderle and Wolfe, 1988). It is more expedient, however, to simply simulate a solution for $x_1(t)$ and then find V_{max} as a function of load.

Using a simulation to reproduce the isotonic experiment, elasticity's estimated from the length-tension curves as previously described and data from rectus eye muscle, parameter values for the viscous elements in the muscle model are found as B_1 = 2.0 Nsm^{-1} and B_2 = 0.5 Nsm^{-1} as demonstrated by Enderle and co-workers in 1991. The viscous element B_1 is estimated from the time constant from the isotonic time course. The viscous element B_2 is calculated by trial and error so that the simulated force-velocity curve matches the experimental force-velocity curve.

Shown in Fig. 5.3 are the force-velocity curves using the model described in Eq. (5.8) (with

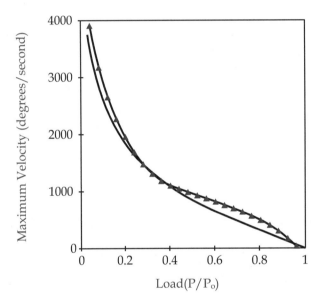

Figure 5.3: Force-velocity curve derived from simulation studies with the linear muscle model with an end stop. Shown with triangles indicating simulation calculation points and an empirical fit to the force-velocity data (solid line) as described by Close and Luff (1974). Adapted from Enderle et al. (1990). A Comparison of Static and Dynamic Characteristics Between Rectus Eye Muscle and Linear Muscle Model Predictions, *IEEE Trans. on Biomedical Engineering*, vol. 38, no. 12, pp. 1235–1245, 1991.

triangles), plotted along with an empirical fit to the data (solid line). It is clear that the force-velocity curve for the linear muscle model is hardly a straight line, and that this curve fits the data well.

The muscle lever model described by Eqs. (5.6) and (5.7) is a third-order linear system and is characterized by three poles. Dependent on the values of the parameters, the eigenvalues (or poles) consist of all real poles or a real and a pair of complex conjugate poles. A real pole is the dominant eigenvalue of the system. Through a sensitivity analysis, viscous element B_1 is the parameter that has the greatest effect on the dominant eigenvalue for this system, while viscous element B_2 has very little effect on the dominant eigenvalue. Thus, viscous element B_1 is estimated so that the dominant time constant of the lever system model (approximately $\frac{B_1}{K_{lt}}$ when $B_1 > B_2$) matches the time constant from the isotonic experimental data. For rectus eye muscle data, the duration of the isotonic experiment is approximately 100 ms. A value for $B_1 = 2.0$ Nsm^{-1} yields a simulated isotonic response with approximately the same duration. For skeletal muscle data, the duration of the isotonic experiment is approximately 400 ms, and a value for $B_1 = 600$ Nsm^{-1} yields a simulated isotonic response with approximately the same duration. It is known that fast and slow muscle have differently shaped force velocity curves and that the fast muscle force-velocity curve data has less curvature. Interestingly, the changes in the parameter values for B_1 as suggested here gives differently shaped force velocity curves consistent with fast (rectus eye muscle) and slow (skeletal muscle) muscle.

The parameter value for viscous element B_2 is selected by trial and error so that the shape of the simulation force-velocity curve matches the data. As the value for B_2 is decreased from 0.5 Nsm^{-1}, the shape of the force velocity curves changes to a more linear shaped function. Moreover, if the value of B_2 falls below approximately 0.3 Nsm^{-1}, strong oscillations appear in the simulations of the isotonic experiment, which are not present in the data. Thus, the viscous element B_2 is an essential component in the muscle model. Without it, the shape of the force-velocity curve is linear and the time course of the isotonic experiment do not match the characteristics of the data.

Varying the parameter values of the lever muscle model changes the eigenvalues of the system. For instance, with $M = 0.5$ kg, the system's nominal eigenvalues (as defined with the parameter values previously specified) are a real pole at -30.71 and a pair of complex conjugate poles at $-283.9 \pm j221.2$. If the value of B_2 is increased, three real eigenvalues describe the system. If the value of B_2 is decreased, a real pole and a pair of complex conjugate poles continue to describe the system. Changing the value of B_1 does not change the eigenvalue composition, but it does significantly change the value of the dominant eigenvalue from -292 with $B_1 = 0.1$ to -10 with $B_1 = 6$.

5.3   1995 LINEAR HOMEOMORPHIC SACCADIC EYE MOVEMENT MODEL

The linear model of the oculomotor plant presented in Section 3.6 is based on a nonlinear oculomotor plant model by Hsu and coworkers using a linearization of the force-velocity relationship and elasticity curves. Using the linear model of muscle described in the previous section, it is possible to avoid the linearization and to derive a truer linear homeomorphic saccadic eye movement model.

The linear muscle model in the previous section has the static and dynamic properties of rectus eye muscle, a model without any nonlinear elements. As presented, the model has a nonlinear force-velocity relationship that matches eye muscle data using linear viscous elements, and the length tension characteristics are also in good agreement with eye muscle data within the operating range of the muscle. Some additional advantages of the linear muscle model are that a passive elasticity is not necessary if the equilibrium point $x_e = -19.3°$, rather than $15°$, and muscle viscosity is a constant that does not depend on the innervation stimulus level.

Figure 5.4 illustrates the mechanical components of the updated oculomotor plant for horizontal eye movements, the lateral and medial rectus muscle, and the eyeball. The agonist muscle is modeled as a parallel combination of viscosity B_2 and series elasticity K_{se}, connected to the parallel combination of active state tension generator F_{ag}, viscosity element B_1 and length tension elastic element K_{lt}. For simplicity, agonist viscosity is set equal to antagonist viscosity. The antagonist muscle is similarly modeled with a suitable change in active state tension to F_{ant}. Each of the elements defined in the oculomotor plant is ideal and linear.

The eyeball is modeled as a sphere with moment of inertia J_p, connected to a pair of viscoelastic elements connected in series. The update of the eyeball model is based on observations by Robinson, presented in 1981 (Robinson, D., 1981), and the following discussion. In the model of the oculomotor plant described in Section 3.6, passive elasticity K_{pe} was combined with the passive elastic orbital tissues. In the new linear model muscle presented in this chapter, the elastic element K_{pe} is no longer included in the muscle model. Thus, the passive orbital tissue elasticity needs to be updated due to the elimination of K_{pe} and the observations by Robinson. As reported by Robinson, D. (1981), "When the human eye, with horizontal recti detached, is displaced and suddenly released, it returns rapidly about 61% of the way with a time constant of about 0.02 sec, and then creeps the rest of the way with a time constant of about 1 sec." As suggested according to this observation, there are at least two viscoelastic elements. Here it is proposed that these two viscoelastic elements replace the single viscoelastic element of the previous oculomotor plant. Connected to the sphere, are $B_{p1} || K_{p1}$ connected in series to $B_{p2} || K_{p2}$. As reported by Robinson, total orbital elasticity is equal to 12.8×10^{-7} g/° (scaled for this model). Thus, with the time constants previously described, the orbital viscoelastic elements are evaluated as $K_{p1} = 1.28 \times 10^{-6}$ g/°, $K_{p2} = 1.98 \times 10^{-6}$ g/°, $B_{p1} = 2.56 \times 10^{-8}$ gs/°, and $B_{p2} = 1.98 \times 10^{-6}$ gs/°. For modeling purposes, θ_5 is the variable associated with the change from equilibrium for these two pairs of viscoelastic elements. Both θ and θ_5 are removed from the analysis for simplicity using the substitution $\theta = 57.296 \frac{x}{r}$ and $\theta_5 = 57.296 \frac{x_5}{r}$.

By summing the forces acting at junctions 2 and 3, and the torques acting on the eyeball and junction 5, a set of four equations is written to describe the oculomotor plant.

$$
\begin{aligned}
F_{ag} &= K_{lt}x_2 + B_1\dot{x}_2 + K_{se}(x_2 - x_1) + B_2(\dot{x}_2 - \dot{x}_1) \\
B_2(\dot{x}_4 - \dot{x}_3) &+ K_{se}(x_4 - x_3) = F_{ant} + K_{lt}x_3 + B_1\dot{x}_3 \\
B_2(\dot{x}_2 + \dot{x}_3 - \dot{x}_1 - \dot{x}_4) &+ K_{se}(x_2 + x_3 - x_1 - x_4) = J\ddot{x} + B_3(\dot{x} - \ddot{y}\dot{x}_5) + K_1(x - x_5) \\
K_1(x - x_5) &+ B_3(\dot{x} - \dot{x}_5) = B_4\dot{x}_5 + K_2x_5
\end{aligned}
\tag{5.9}
$$

where

$$
J = \frac{57.296}{r^2}J_p, \quad B_3 = \frac{57.296}{r^2}B_{p1}, \quad B_4 = \frac{57.296}{r^2}B_{p2}, \quad K_1 = \frac{57.296}{r^2}K_{p1}, \quad K_2 = \frac{57.296}{r^2}K_{p2}.
$$

Using Laplace variable analysis about an operating point as before in Section 3.6, yields

$$
\begin{aligned}
K_{se}K_{12}(F_{ag} - F_{ant}) &+ (K_{se}B_{34} + B_2K_{12})(\dot{F}_{ag} - \dot{F}_{ant}) + B_2B_{34}(\ddot{F}_{ag} - \ddot{F}_{ant}) \\
&= C_4\dddot{x} + C_3\dddot{x} + C_2\ddot{x} + C_1\dot{x} + C_0x
\end{aligned}
\tag{5.10}
$$

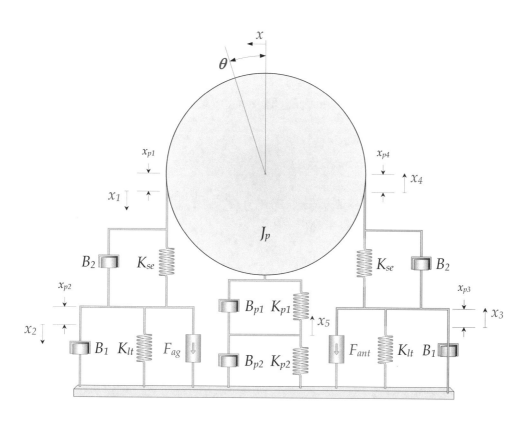

Figure 5.4: This diagram illustrates the mechanical components of the updated oculomotor plant. The muscles are shown to be extended from equilibrium, a position of rest, at the primary position (looking straight ahead), consistent with physiological evidence. The average length of the rectus muscle at the primary position is approximately 40 mm, and at the equilibrium position is approximately 37 mm. θ is the angle the eyeball is deviated from the primary position, and variable x is the length of arc traversed. When the eye is at the primary position, both θ and x are equal to zero. Variables x_1 through x_4 are the displacements from equilibrium for the stiffness elements in each muscle, and θ_5 is the rotational displacement for passive orbital tissues. Values x_{p1} through x_{p4} are the displacements from equilibrium for each of the variables x_1 through x_4 at the primary position. The total extension of the muscle from equilibrium at the primary position is x_{p1} plus x_{p2} or x_{p3} plus x_{p4}, which equals approximately 3 mm. It is assumed that the lateral and medial rectus muscles are identical, such that x_{p1} equals x_{p4} and x_{p3} equals x_{p2}. The radius of the eyeball is r.

where

$$B_{12} = B_1 + B_2, \quad B_{34} = B_3 + B_4, \quad K_{12} = K_1 + K_2$$
$$C_4 = J B_{12} B_{34}$$
$$C_3 = B_3 B_4 B_{12} + 2 B_1 B_2 B_{34} + J B_{34} K_{st} + J B_{12} K_{12}$$
$$C_2 = 2 B_1 B_{34} K_{se} + J K_{st} K_{12} + B_3 B_{34} K_{st} + B_3 B_{12} K_{12} + K_1 B_{12} B_{34} - B_3^2 K_{st} - 2 K_1 B_3 B_{12}$$
$$\quad + 2 B_2 K_{lt} B_{34} + 2 B_1 K_{12} B_2$$
$$C_1 = 2 K_{lt} B_{34} K_{se} + 2 B_1 K_{12} K_{se} + B_3 K_{st} K_2 + K_1 B_{34} K_{st} + K_1 B_{12} K_{12} - K_{st} K_1 B_3$$
$$\quad - K_1^2 B_{12} + 2 B_2 K_{lt} K_{12}$$
$$C_0 = 2 K_{lt} K_{se} K_{12} + K_1 K_{st} K_2.$$

Converting from x to θ gives

$$\delta \left(K_{se} K_{12} \left(F_{ag} - F_{ant} \right) + \left(K_{se} B_{34} + B_2 K_{12} \right) \left(\dot{F}_{ag} - \dot{F}_{ant} \right) + B_2 B_{34} \left(\ddot{F}_{ag} - \ddot{F}_{ant} \right) \right)$$
$$= \dddot{\theta} + P_3 \dddot{\theta} + P_2 \ddot{\theta} + P_1 \dot{\theta} + P_0 \theta \qquad (5.11)$$

where

$$\delta = \frac{57.296}{r J B_{12} B_{34}}, \quad P_3 = \frac{C_3}{C_4}, \quad P_2 = \frac{C_2}{C_4}, \quad P_1 = \frac{C_1}{C_4}, \quad P_0 = \frac{C_0}{C_4}.$$

Based on an analysis of experimental data, suitable parameter estimates for the oculomotor plant are:

$$K_{SE} = 125 \ \mathrm{Nm}^{-1}$$
$$K_{LT} = 60.7 \ \mathrm{Nm}^{-1}$$
$$B_1 = 2.0 \ \mathrm{Nsm}^{-1}$$
$$B_2 = 0.5 \ \mathrm{Nsm}^{-1}$$
$$J = 2.2 \times 10^{-3} \ \mathrm{Ns^2 m}^{-1}$$
$$B_3 = 0.538 \ \mathrm{Nsm}^{-1}$$
$$B_4 = 41.54 \ \mathrm{Nsm}^{-1}$$
$$K_1 = 26.9 \ \mathrm{Nm}^{-1}$$
$$K_2 = 41.54 \ \mathrm{Nm}^{-1}$$

Based on the updated model of muscle and length tension data presented in the previous section, steady-state active state tensions are determined as:

$$F = \begin{cases} 0.4 + 0.0175\theta & \mathrm{N} \quad \text{for} \quad \theta \geq 0° \\ 0.4 + 0.0125\theta & \mathrm{N} \quad \text{for} \quad \theta < 0°. \end{cases} \qquad (5.12)$$

The agonist and antagonist active state tensions follow from Fig. 3.5, which assume no latent period, and are given by the following low-pass filtered waveforms:

$$\dot{F}_{ag} = \frac{N_{ag} - F_{ag}}{\tau_{ag}} \quad \text{and} \quad \dot{F}_{ant} = \frac{N_{ant} - F_{ant}}{\tau_{ant}} \qquad (5.13)$$

where N_{ag} and N_{ant} are the neural control inputs (pulse-step waveforms), and

$$\tau_{ag} = \tau_{ac}(u(t) - u(t - t_1)) + \tau_{de}u(t - t_1)$$
$$\tau_{ant} = \tau_{de}(u(t) - u(t - t_1)) + \tau_{ac}u(t - t_1)$$

are the time-varying time constants.

Saccadic eye movements simulated with this model have characteristics that are in good agreement with the data, including position, velocity and acceleration, and the main sequence diagrams. As before, the relationship between agonist pulse magnitude and pulse duration is tightly coupled.

Example 5.1   Using the oculomotor plant model described with Eq. (5.11), parameters given in this section and the steady-state input from Eq. (5.12), simulate a 10° saccade. Plot agonist and antagonist active-state tension, position, velocity and acceleration vs. time. Compare the simulation with the main sequence diagram in Fig. 1.5.

Solution.
The solution to this example involves selecting a set of parameters $\left(F_p,\ t_1,\ \tau_{ac},\ \text{and}\ \tau_{de}\right)$ that match the characteristics observed in the main sequence diagram shown in Fig. 1.5. There is a great deal of flexibility in simulating a 10° saccade. The only constraints for the 10° saccade simulation results are that the duration is approximately 40 to 50 ms and peak velocity in the 500 to 600°s^{-1} range. For realism, a latent period of 150 ms has been added to the simulation results. A SIMULINK block diagram of Eq. (5.11) is shown in Fig. 5.5.

The response of the system is shown in Fig. 5.6 with $F_p = 1.3$ N, $t_1 = 0.01$ s, $\tau_{ac} = 0.018$ s and $\tau_{de} = 0.018$ s. These 10° simulation results have the main sequence characteristics with a peak velocity of 568°s^{-1} and a duration of 45 ms.

Many other parameter sets can also simulate a 10° saccade. For instance, consider reducing τ_{de} to .009 s. Because the antagonist active-state tension activity goes toward zero more quickly than in the last case, a greater total active-state tension $\left(F_{ag} - F_{ant}\right)$ results. Therefore, to arrive at 10° with the appropriate main sequence characteristics, F_p needs to be reduced to 1.0 N if τ_{ac} remains at 0.018 s and t_1 equals 0.0115 s. This 10° simulation is shown Fig. 5.7.

To simulate larger saccades with main sequence characteristics, the time constants for the agonist and antagonist active-state tensions can be kept at the same values as the 10° saccades or made functions of saccade amplitude (see Bahill, A., 1981 for several examples of amplitude dependent time constants). Main sequence simulations for 15° and 20° saccades are obtained with $F_p = 1.3$ N and the time constants are both fixed at 0.018 s (the first case) by changing t_1 to 0.0155 and 0.0223 s, respectively. For example, the 20° simulation results are shown in Fig. 5.8 with a peak velocity of 682°s^{-1} and a duration of 60 ms.

In general, as F_p increases, t_1 decreases to maintain the same saccade amplitude. Additionally, peak velocity increases as F_p increases. For the saccade amplitude to remain a constant as either or both time constant increases, F_p should also increase.

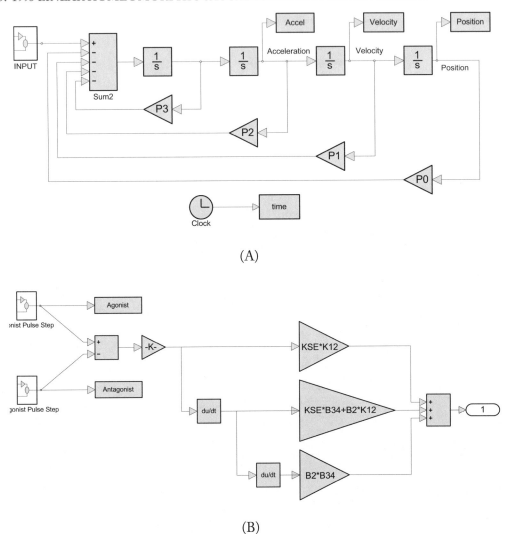

Figure 5.5: SIMULINK block diagrams for Example 5.1. (A) Model defined by Eq. (5.11). (B) The input.

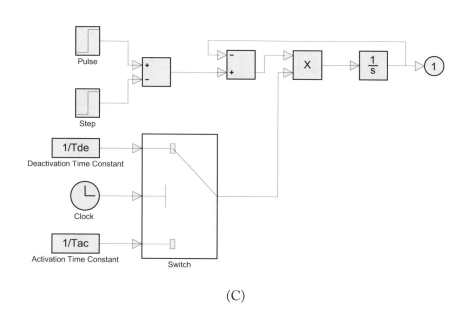

(C)

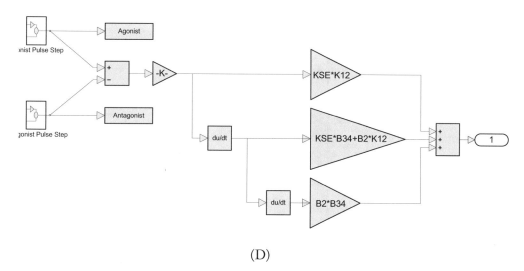

(D)

Figure 5.5: (continued). SIMULINK block diagrams for Example 5.1. (C) Agonist Pulse-Step. (D) Antagonist Pulse-Step.

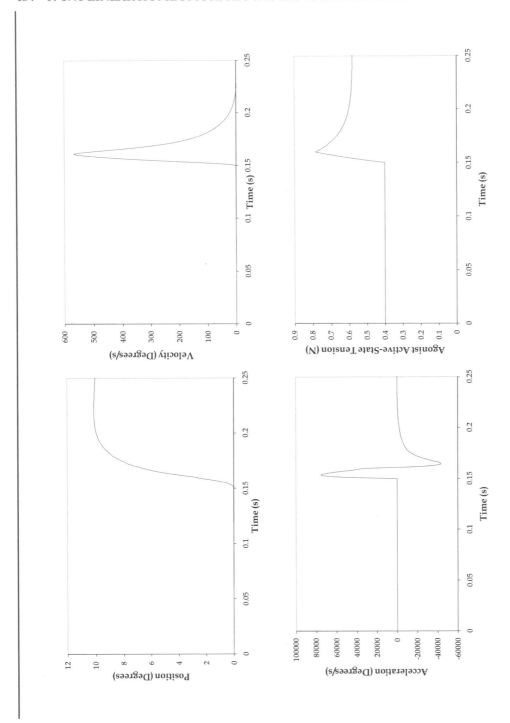

Figure 5.6: Simulation for Example 5.1 With $F_p = 1.3$ N, $t_1 = 0.10$ s, $\tau_{ac} = 0.018$ s, and $\tau_{de} = 0.018$ s.

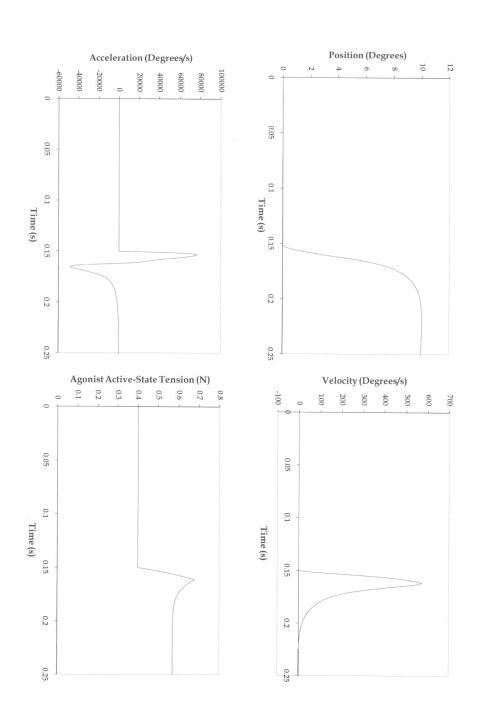

Figure 5.7: Simulation for Example 5.1 With $F_p = 1.0$ N, $t_1 = 0.010$ s, $\tau_{ac} = 0.018$ s, and $\tau_{de} = 0.009$ s.

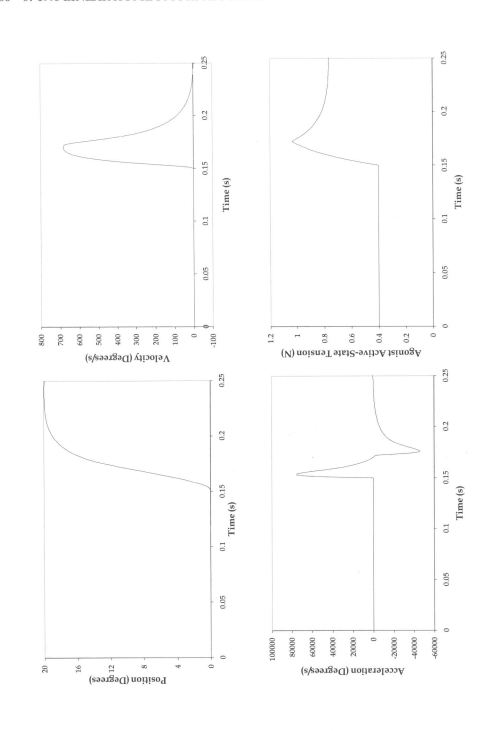

Figure 5.8: Simulation for Example 5.1 With $F_p = 1.3$ N, $t_1 = 0.0223$ s, $\tau_{ac} = 0.018$ s, and $\tau_{de} = 0.018$ s.

Bibliography

Albano, J.E. and Wurtz, R.H., (1981). The role of the primate superior colliculus, pretectum, and posterior-medial thalamus in visually guided eye movements, *Progress in Oculomotor Research*, (edited by Fuchs and Becker). Elsevier, North Holland, pp. 145–153.

Anderson R.W., Keller, E.L., Gandhi, N.J., Das, S., (1998). Two dimensional saccade-related population activity in superior colliculus in monkey. *J Neurophysiol* 80: 798–817.

Bahill, A.T., Clark, M.R. and Stark, L., (1975). The Main Sequence, A Tool For Studying Human Eye Movements, *Math. Biosci.*, 24: 194–204. DOI: 10.1016/0025-5564(75)90075-9 4

Bahill, A.T. and Hamm, T.M., (1989). Using open-loop experiments to study physiological systems, with examples from the human eye movement systems, *News in Physiol. Sci.*, 4: 104–109. 8, 22, 25

Bahill, A.T. and Harvey, D.R., (1986). Open-loop experiments for modeling the human eye movement system, *IEEE Trans. Sys. Man. Cybern.*, SMC-16(2): 240–250. DOI: 10.1109/TSMC.1986.4308944 8, 22

Bahill, A.T. and McDonald, J.D., (1983). Model emulates human smooth pursuit system producing zero-latency target tracking, *Biol. Cyber.*, 48: 213–222. DOI: 10.1007/BF00318089 8, 22

Bahill, A.T. and McDonald, J.D., (1983b). Frequency limitations and optimal step size for the two-point central difference derivative algorithm with applications to human eye movement data, *IEEE Trans. Biomed. Eng.*, BME-30(3): 191–194. DOI: 10.1109/TBME.1983.325108 110

Bahill, A.T., *Bioengineering: Biomedical, Medical and Clinical Engineering*, Prentice-Hall, Englewood Cliffs, NJ, 1981. 93, 101, 102, 131

Bahill, A.T., Brockenbrough A., and Troost, B.T., (1981). Variability and development of a normative data base for saccadic eye movements, *Invest. Ophthal. uis. Sci.*, 21, 116–125. 77

Bahill, A.T., Latimer, J.R., and Troost, B.T., (1980). Linear homeomorphic model for human movement, *IEEE Trans. Biomed Engr.*, BME-27, 631–639. DOI: 10.1109/TBME.1980.326703 22, 62, 66, 75, 77, 79, 101

Bahill, A.T., (1980). "Development, validation and sensitivity analyses of human eye movement models," *CRC Crit. Rev. Bioeng.*, vol. 4, no. 4, pp. 311–355. 57, 92

Bahill, A.T, Clark, M.R., Stark, L., (1975b). Dynamic overshoot in saccadic eye movements is caused by neurological control signal reversals. *Exp Neurol*, 48: 107–122. DOI: 10.1016/0014-4886(75)90226-5

Becker, W. and Fuchs, A.F., (1985). Prediction in the oculomotor system: smooth pursuit during transient disappearance of visual target. *Experimental Brain Research*, 57: 562–575. DOI: 10.1007/BF00237843 8

Behan, M. and Kime, N.M., (1996). Intrinsic Circuitry in the Deep Layers of the Cat Superior Colliculus, *Vis. Neurosci.*, 13: 1031–1042. DOI: 10.1017/S0952523800007689

Bruce, C.J., Goldberg, M.E., Bushnell, M.C., and Stanton, G.B., (1985). Primate frontal eye fields. II. Physiological and anatomical correlates of electrically evoked eye movements. *Journal of Neurophysiology*, vol. 54, no. 3: 714–34.

Bruce, C.J. and Goldberg, M.E., (1985). Primate frontal eye fields, I. single neurons discharging before saccades, *J. Neurophys.*, 53(3): 603–635.

Buttner-Ennever, J.A., Horn, A.K., Henn, V., and Cohen, B., (1999). Projections from the superior colliculus motor map to omnipause neurons in monkey, *J. Comp. Neurol.*, Oct 11; 413(1): 55–67. DOI: 10.1002/(SICI)1096-9861(19991011)413:1%3C55::AID-CNE3%3E3.0.CO;2-K

Cannon, S.C. and Robinson, D.A., (1987). Loss of the neural integrator of the oculomotor system from brain stem lesions in monkey, *J. Neurophys.*, 57(5): 1383–1409.

Cannon, S.C. and Robinson, D.A., (1985). An improved neural-network model for the neural integrator of the oculomotor system: more realistic neuron behavior, *Biol. Cyber.*, 53: 93–108. DOI: 10.1007/BF00337026

Cannon, S.C., Robinson, D.A. and Shamma, S., (1983). A proposed neural network for the integrator of the oculomotor system, *Biol. Cyber.*, 49: 127–136. DOI: 10.1007/BF00320393

Clark, M.R. and Stark, L., (1975). Time optimal control for human saccadic eye movement, *IEEE Trans. Automat. Contr.*, AC-20: 345–348. DOI: 10.1109/TAC.1975.1100955 93, 101, 102, 103

Carpenter, R.H.S., *Movements of the Eyes*, 2nd ed., Pion, London, 1988. 10, 73

Chen-Harris, H., Joiner, W.M., Ethier, V., Zee, D.S., and Shadmehr, R., (2008). Adaptive Control of Saccades via internal Feedback, *J. Neurophys.*, 28(11): 2804–2813. DOI: 10.1523/JNEUROSCI.5300-07.2008

Chimoto, S., Iwamoto, Y., Shimazu, H. and Yoshida, K., (1996). Functional connectivity of the superior colliculus with saccade-related brain stem neurons in the cat, *Prog. Brain Res.*, 112: 157–65. DOI: 10.1016/S0079-6123(08)63327-0

Close, M.R. and Luff, A.R., (1974). Dynamic properties of inferior rectus muscle of the rat, *J. Physiol.*, London, 236: 258. 125, 126

Collins, C.C., O'Meara, D. and Scott, A.B., (1975). Muscle tension during unrestrained human eye movements, *J. Physiol.*, 245: 351–369. 43, 44, 80, 92

Collins, C.C., The human oculomotor control systems, *Basic Mechanisms of Ocular Motility and their Clinical Implications*. G. Lennerstrand and P. Bach-y-Rita, Eds., pp. 145–180. Pergamon Press, Oxford, 1975. 46, 79, 121, 123

Contreras, D., Destexhe, D., Sejnowski, T. and Steriade, M., (1997). Spatiotemporal Patterns of Spindle Oscillations in Cortex and Thalamus. *J Neurophysiol*, 17(3):1179–1196.

Cook, G. and Stark, L., (1968). The human eye movement mechanism: experiments, modeling and model testing. *Archs Ophthal.*, 79, pp 428–436. 79

Cook, G. and Stark, L., (1967). Derivation of a model for the human eye positioning mechanism, *Bull. Math. Biophys*, 29, 153–174. DOI: 10.1007/BF02476968 79

Das, S., Gandhi, N.J., and Keller, E.L., (1995). Open-loop simulations of the primate saccadic system using burst cell discharge from the superior colliculus. *Biol. Cybern*, 73:509–518. DOI: 10.1007/BF00199543

Dean, P., (1995). Modelling the role of the cerebellar fastigial nuclei in producing accurate saccades: the importance of burst timing, *Neuroscience*, 68(4): 1059–1077. DOI: 10.1016/0306-4522(95)00239-F

Descartes, R., *Treatise of Man*. Originally published by Carles Angot, Paris, (1664.) Published translation and commentary by T.S. Hall, Harvard University Press, Cambridge, MA, 1972. 31

Destexhe, A., Bal, T., McCormick, D.A., Sejnowski, T.J., (1996). Ionic mechanisms underlying synchronized oscillations and propagating waves in a model of ferret thalamic slices. *J Neurophysiol.*, 76:2049 –2070.

Enderle, J.D., *Eye Movements*. In Wiley Encyclopedia of Biomedical Engineering (Metin Akay, Ed.), Hoboken: John Wiley & Sons, 2006. 4

Enderle, J.D., *The Fast Eye Movement Control System*. In: The Biomedical Engineering Handbook, Biomedical Engineering Fundamentals, 3rd ed., J. Bronzino, Ed., CRC Press, Boca Raton, FL, (2006), Chapter 16, pages 16–1 to 16–21.

Enderle, J.D., Neural Control of Saccades. In J. Hyönä, D. Munoz, W. Heide and R. Radach (Eds.), *The Brain's Eyes: Neurobiological and Clinical Aspects to Oculomotor Research, Progress in Brain Research*, V. 140, Elsevier, Amsterdam, 21–50, 2002. 4

Enderle, J.D., Blanchard, S.M., and Bronzino, J.D., *Introduction to Biomedical Engineering*. Academic Press, San Diego, California, 1062 pages, 2000.

Enderle, J.D. and Engelken, E.J., (1996). Effects of Cerebellar Lesions on Saccade Simulations, *Biomed. Sci. Instru.*, 32: 13–22.

Enderle, J.D. and Engelken, E.J. (1995). Simulation of Oculomotor Post-Inhibitory Rebound Burst Firing Using a Hodgkin-Huxley Model of a Neuron. *Biomedical Sciences Instrumentation*, 31: 53–58.

Enderle, J.D., Engelken, E.J., and Stiles, R.N. (1991). A comparison of static and dynamic characteristics between rectus eye muscle and linear muscle model predictions. *IEEE Trans. Biomed. Eng.* 38:1235–1245.

Enderle, J.D., Engelken, E.J., and Stiles, R.N., (1990). Additional Developments in Oculomotor Plant Modeling. *Biomedical Sciences Instrumentation*, 26: 59–66. 126

Enderle, J.D. and Wolfe, J.W., (1988). Frequency Response Analysis of Human Saccadic Eye Movements: Estimation of Stochastic Muscle Forces, *Comp. Bio. Med.*, 18: 195–219. DOI: 10.1016/0010-4825(88)90046-7 4, 73, 103, 126

Enderle, J.D., (1988). Observations on Pilot Neurosensory Control Performance During Saccadic Eye Movements, *Aviat., Space, Environ. Med.*, 59: 309–313. 7

Enderle, J.D., and Wolfe, J.W., (1987). Time-Optimal Control of Saccadic Eye Movements. *IEEE Transactions on Biomedical Engineering*, Vol. BME-34, No. 1: 43–55. DOI: 10.1109/TBME.1987.326014 4, 98

Enderle, J.D., Wolfe, J.W., and Yates, J.T. (1984). The Linear Homeomorphic Saccadic Eye Movement Model—A Modification. *IEEE Transactions on Biomedical Engineering*, 1984. Vol. BME-31, No. 11: 711–820. 66, 77

Engelken, E.J., Stevens, K.W., McQueen, W.J. and Enderle, J.D., (1996). Application of Robust Data Processing Methods to the Analysis of Eye Movements, *Biomed. Sci. Instru.*, 32: 7–12. 119

Engelken, E.J., Stevens, K.W., Bell, A.F. and Enderle, J.D., (1993). Linear Systems Analysis of the Vestibulo-Ocular Reflex: Clinical Applications, *Biomed. Sci. Instru.*, 29: 319–326. 119

Engelken, E.J., Stevens, K.W. and Enderle, J.D., (1991). Optimization of an Adaptive Non-linear Filter for the Analysis of Nystagmus, *Biomed. Sci. Instru.*, 27: 163–170. 119

Engelken, E.J., Stevens, K.W. and Enderle, J.D., (1991a). Relationships between Manual Reaction Time and Saccade Latency in Response to Visual and Auditory Stimuli, *Aviat., Space, Environ. Med.*, 62: 315–318.

Engelken, E.J., Stevens, K.W. and Enderle, J.D., (1990). Development of a Non-Linear Smoothing Filter for the Processing of Eye-Movement Signals, *Biomed. Sci. Instru.*, 26: 5–10. 119

Engelken, E.J. and Stevens, K.W., (1989). Saccadic Eye Movements in Response to Visual, Auditory, and Bisensory Stimuli, *Aviat., Space, Environ. Med.*, 762–768.

Engelken, E.J., Stevens, K.W., Wolfe, J.W., and Yates, J.T., (1984). A Limbus Sensing Eye-Movement Recorder. Brooks AFB, TX: USAF School of Aerospace Medicine. *USAFSAM-TR-84–29.* 5

Fenn, W.O. and Marsh, B.S., (1935). Muscular force a different speeds of shortening, *J. Physiol.*, London, 85: 277–297. 75, 125

Fuchs, A.F., Kaneko, C.R.S., Scudder, C.A. (1985). Brainstem Control of Saccadic Eye Movements. *Ann Rev. Neurosci*, 8:307–337. DOI: 10.1146/annurev.ne.08.030185.001515

Fuchs, A.F. and Luschei, E.S., (1971). Development of isometric tension in simian extraocular muscle. *J. Physiol.*, 219, 155–66.

Fuchs, A.F. and Luschei, E.S., (1970). Firing patterns of abducens neurons of alert monkeys in relationship to horizontal eye movement. *J. Neurophysiol.*, 33 (3), 382–392.

Galiana, H.L., (1991). A Nystagmus strategy to linearize the vestibulo-ocular reflex, *IEEE Trans. Biomed. Eng.*, 38: 532–543. DOI: 10.1109/10.81578

Gancarz, G. and Grossberg, S., (1998). A neural model of the saccade generator in reticular formation. *Neural Networks*, vol. 11: pp. 1159–1174. DOI: 10.1016/S0893-6080(98)00096-3

Gandhi, N.J. and Keller, E.L., (1997). Spatial distribution and discharge characteristics of the superior colliculus neurons antidromically activated from the omnipause region in monkey. *J. Neurophysiol.*, 76: 2221–5.

Girard, B. and Berthoz, A., (2005). From brainstem to cortex: Computational models of saccade generation circuitry. *Progress in Neurobiology*, 77: 215–251. DOI: 10.1016/j.pneurobio.2005.11.001

Goffart, L. and Pelisson, D., (1997). Changes in initiation of orienting gaze shifts after muscimol inactivation of the caudal fastigial nucleus in the cat, *J. Physiology*, Sept 15, 503(3): 657–71. DOI: 10.1111/j.1469-7793.1997.657bg.x

Goldberg, M.E. and Bushnell, M.C., (1981). Behavioral enhancement of visual responses in monkey cerebral cortex. II. Modulation in frontal eye fields specifically related to saccades. *Journal of Neurophysiology*, vol. 46, no. 4: 773–87.

Goldstein, H.P. and Robinson, D.A., (1986). Hysteresis and slow drift in abducens unit activity. *Journal of Neurophysiology*, vol. 55, pp. 1044–1056.

Goldstein, H. and Robinson, D., (1984). A two-element oculomotor plant model resolves problems inherent in a single-element plant model. *Society for Neuroscience Abstracts, 10,* 909.

Goldstein, H., (1983). The neural encoding of saccades in the rhesus monkey (Ph.D. dissertation). Baltimore, MD: The Johns Hopkins University.

Graybiel, A.M., (1978). Organization of the nigrotectal connection: an experimental tracer study in the cat. *Brain Research,* vol. 143, no. 2: 339–48. DOI: 10.1016/0006-8993(78)90573-5

Harris, C.M. and Wolpert, D.M., (2006). The main sequence of saccades optimizes speed-accuracy trade-off. *Biol Cybern,* 95 (1), 21–29. DOI: 10.1007/s00422-006-0064-x

Harting J.K., (1977). Descending pathways from the superior colliculus: an autoradiographic analysis in the rhesus monkey (Macaca mulatta). *J. Comp. Neurol.,* 173: 583–612. DOI: 10.1002/cne.901730311

Harwood, M.R., Mezey, L.E., and Harris, C.M., (1999). The Spectral Main Sequence of Human Saccades, *The Journal of Neuroscience,* 19(20): 9098–9106. 4

Hashimoto, M. and Ohtsuka, K., (1995). Transcranial magnetic stimulation over the posterior cerebellum during visually guided saccades in man, *Brain,* 118,(5): 1185–93. DOI: 10.1093/brain/118.5.1185

Hikosaka, O. and Wurtz, R.H., The role of substantia nigra in the initiation of saccadic eye movements, *Progress in Oculomotor Research,* (edited by Fuchs and Becker), Elsevier, North Holland, pp. 145–153, 1981.

Hikosaka, O. and Wurtz, R.H., (1983a). Visual and oculomotor functions of monkey substantia nigra pars reticulata. I. Relation of visual and auditory responses to saccades, *J. Neurophys.,* May 49(5): 1230–53.

Hikosaka, O. and Wurtz, R.H., (1983b). Visual and oculomotor functions of monkey substantia nigra pars reticulata. II. Visual responses related to fixation of gaze, *J. Neurophys.,* May 49(5): 1254–67.

Hill, A.V., (1951a). The transition from rest to full activity in muscles: the velocity of shortening, *Pro. Royal Soc.,* London (B), 138: 329–338. DOI: 10.1098/rspb.1951.0026

Hill, A.V., (1951b). The effect of series compliance on the tension developed in a muscle twitch, *Pro. Royal Soc.,* London (B), 138: 325–329. DOI: 10.1098/rspb.1951.0025

Hill, A.V., (1950a). The development of the active state of muscle during the latent period, *Pro. Royal Soc.,* London (B), 137: 320–329. DOI: 10.1098/rspb.1950.0043

Hill, A.V., (1950b). The series elastic component of muscle, *Pro. Royal Soc.,* London (B), 137: 273–280. DOI: 10.1098/rspb.1950.0035

Hill, A.V., (1938). The heat of shortening and dynamic constants of muscle, *Pro. Royal Soc.*, London (B), 126: 136–195. DOI: 10.1098/rspb.1938.0050 75

Hodgkin, A.L., Huxley, A.F., and Katz, B., (1952). Measurement of Current-Voltage Relations in the Membrane of the Giant Axon of *Loligo*. *Journal of Physiology*, 116: 424–448.

Horwitz, G.D. and Newsome, W.T., (1999). Separate signals for target selection and movement specification in the superior colliculus, *Science,* May 14; 284(5417): 1158–61. DOI: 10.1126/science.284.5417.1158

Hung, G.K. and Ciuffreda, K.J., *Models of the Visual System*. Kluwer Academic/Plenum Publishers, New York, NY, 2002.

Hsu, F.K., Bahill, A.T. and Stark, L., (1976). Parametric sensitivity of a homeomorphic model for saccadic and vergence eye movements, *Comp. Prog. Biomed.*, 6: 108–116. DOI: 10.1016/0010-468X(76)90032-5 31, 57, 59, 77, 79, 103

Hu, X., Jiang, H., Gu, C., Li, C. and Sparks, D. (2007). Reliability of Oculomotor Command Signals Carried by Individual Neurons. *PNAS*, 8137–8142. DOI: 10.1073/pnas.0702799104

Ito, M., (1984). The modifiable neuronal network of the cerebellum, *Jpn. J. of Physiol.*, 34: 781–792. DOI: 10.2170/jjphysiol.34.781

Jahnsen, H. and Llinas, R., (1984a). Electrophysiological properties of guinea pig thalamic neurones: an in vitro study, *J. Physiol.,* London, 349: 205–226.

Jahnsen, H. and Llinas, R., (1984b). Ionic basis for the electroresponsiveness and oscillatory properties of guinea pig thalamic neurons in vitro, *J. Physiol.,* London, 349: 227–247.

Kandel, E.R., Schwartz, J.H., and Jessell, T.M., *Principles of Neural Science: Fourth Edition*. McGraw-Hill, New York, 2000.

Kapoula, Z., Robinson, D.A. and Hain, T.C., (1986). Motion of the eye immediately after a saccade. *Exp Brain Res*, 61: 386–394. DOI: 10.1007/BF00239527

Keller, E.L., McPeek, R.M. and Salz, T., (2000). Evidence against direct connections to PPRF EBNs from SC in the monkey, *J. Neurophys.,* 84(3): 1303–13.

Keller, E.L., The Brainstem. In: *Eye Movements*, edited by Carpenter RHS. London: Macmillan, pp. 200–223, 1991.

Keller, E.L., (1989). The cerebellum, *Rev. Oculo. Res.*, 3: 391–411.

Kinariwala, B.K., (1961). Analysis of time varying networks, *IRE Int. Convention Rec.*, 4: 268–276. 96

Krauzlis, R.J., (2005). The control of voluntary eye movements: new perspectives. *The Neuroscientist*, 11 (2),124–137. DOI: 10.1177/1073858404271196

Krauzlis, R.J. and Miles, F.A., (1998). Role of the oculomotor vermis in generating pursuit and saccades: effects of microstimulation, *J. Neurophysiolgy*, Oct 80(4): 2046–62.

LaCroix, T.P., Enderle, J.D. and Engelken, E.J., (1990). Characteristics of Saccadic Eye Movements Induced by Auditory Stimuli, *Biomed. Sci. Instru.*, 26: 67–78.

Lee, C., Roher, W.H. and Sparks, D.L., (1988). Population coding of saccadic eye movements by neurons in the superior colliculus, *Nature*, 332(24): 357–360. DOI: 10.1038/332357a0

Lee, E.B., (1960). Mathematical aspects of the synthesis of linear, minimum response-time controllers, *IRE Trans. Automat. Contr.*, AC-5: 283–289. DOI: 10.1109/TAC.1960.1105031 96

Lefevre, P., Quaia, C., and Optican, L.M., (1998). Distributed model of control of saccades by superior colliculus and cerebellum. *Neural Networks*, vol. 11: pp. 1175–1190. DOI: 10.1016/S0893-6080(98)00071-9

Leichnetz, G.R. and Gonzalo-Ruiz, A., (1996). Prearcuate cortex in the Cebus monkey has cortical and subcortical connections like the macaque frontal eye field and projects to fastigial-recipient oculomotor-related brainstem nuclei [published erratum appears in Brain Res. Bull. 1997; 42(1): following III], *Brain Res. Bull.*, 41(1): 1–29. DOI: 10.1016/0361-9230(96)00154-2

Leigh, R.J. and Zee, D.S., *The Neurology of Eye Movements*. Oxford University Press, New York, NY, 1999. 8

Lehman, S. and Stark, L., (1982). Three algorithms for interpreting models consisting of ordinary differential equations: sensitivity coefficients, sensitivity functions, global optimization, *Math. Biosci*, 62, 107–122. DOI: 10.1016/0025-5564(82)90064-5 77, 78

Lehman, S. and Stark, L., (1979). Simulation of linear and nonlinear eye movement models: sensitivity analyses and enumeration studies of time optimal control, *J. Cybernet. Inf Sci*, 2, 21–43. 77, 93, 95, 102

Ling, L., Fuchs, A., Siebold, C. and Dean, P., (2007). Effects of initial eye position on saccade-related behavior of abducens nucleus neurons in the primate. *J. Neurophysiol.*, 98 (6), 3581–3599. DOI: 10.1152/jn.00992.2007

Markham, C.H., (1981). Cat medial pontine neurons in vestibular nystagmus, *Annals New York Academy of Science*, 374: 189–209. DOI: 10.1111/j.1749-6632.1981.tb30870.x

May, A., (1985). An Improved Human Oculomotor Model For Horizontal Saccadic (Fast) Eye Movements, MS. Thesis, North Dakota State University, Fargo, ND. 79

Meredith, M.A. and Ramoa, A.S., (1998). Intrinsic circuitry of the superior colliculus: pharmaco-physiological identification of horizontally oriented inhibitory interneurons, *J. Neurophys.*, Mar 79(3): 1597–602.

Miura, K. and Optican, L., (2006). Membrane Chanel Properties of Premotor Excitatory Burst Neurons May Underlie Saccade Slowing After lesions of Ominpause Neurons. *J. Comput. Neurosci.*, 20: 25–41. DOI: 10.1007/s10827-006-4258-y

Moschovakis, A.K., Scudder, C.A. and Highstein, S.M., (1996). The Microscopic Anatomy and Physiology of the Mammalian Saccadic System, *Prog. Neurobiol.*, Oct 50(2–3): 133-254. DOI: 10.1016/S0301-0082(96)00034-2

Munoz, D.P. and Istvan, P.J., (1998). Lateral inhibitory interactions in the intermediate layers of the monkey superior colliculus, *J. Neurophys.*, Mar 79(3): 1193–209.

Munoz, D.P., Wurtz, R.H., (1995a). Saccade-related activity in monkey superior colliculus. I. Characteristics of burst and buildup cells. *J. Neurophysiol.,* 73: 2313–2333.

Munoz, D.P. and Wurtz, R.H., (1995b). Saccade-related activity in monkey superior colliculus. II. Spread of activity during saccades. *J. Neurophysiol.,* 73: 2334–2348.

Munoz, D.P. and Wurtz, R.H., (1993a). Fixation cells in the monkey superior colliculus. I. Characteristics of cell discharge, *J. Neurophys.,* 70(2): 559–575.

Munoz, D.P. and Wurtz, R.H., (1993b). Fixation cells in the monkey superior colliculus. II. Reversible activation and deactivation, *J. Neurophys.,* 70(2): 576–589.

Munoz, D.P. and Wurtz, R.H., (1992). Role of the Rostral Superior Colliculus in Active Visual Fixation and Execution of Express Saccades, *J. Neurophys.,* 67(4): 1000–1002.

Munoz, D.P., Pelisson, D. and Guitton, D., (1991). Movement of Neural Activity on the Superior Colliculus Motor Map During Gaze Shifts, *Science,* 251(4999): 1358–60. DOI: 10.1126/science.2003221

Nakahara, H., Morita, K., Wurtz, R.H., and Optican, L.M., (2006). Saccade-Related Spread of Activity Across Superior Colliculus May Arise From Asymmetry of Internal Connections. *J. Neurophysiol.,* 96: 765–774. DOI: 10.1152/jn.01372.2005

Ohki, Y., Shimazu, H., and Suzuki, I., (1988). Excitatory input to burst neurons from the labyrinth and its mediating pathway in the cat: location and functional characteristics of burster-driving neurons. *Exp. Brain Res.,* 72: 457–472. DOI: 10.1007/BF00250591

Ohtsuka, K. and Nagasaka, Y., (1999). Divergent axon collaterals from the rostral superior colliculus to the pretectal accommodation-related areas and the omnipause neuron area in the cat, *J. Comp. Neurol.,* 11; 413(1): 68–76,. DOI: 10.1002/(SICI)1096-9861(19991011)413:1%3C68::AID-CNE4%3E3.0.CO;2-7

Olivier, E., Grantyn, A., Chat, M. and Berthoz, A., (1993). The control of slow orienting eye movements by tectoreticulospinal neurons in the catbehavior, discharge patterns and underlying connections. *Exp. Brain Res.*, 93: 435–449. DOI: 10.1007/BF00229359

Optican, L.M. and Miles, F.A., (1985). Visually induced adaptive changes in primate saccadic oculomotor control signals. *J. Neurophysiol.*, 54 (4), 940–958.

Optican, L.M. and Miles, F.A., (1980). Cerebellar-dependent adaptive control of primate saccadic system, *J. Neurophys.*, 44(6): 1058–1076.

Ottes, F.P., Van Gisbergen, J.A.M. and Eggermont, J.J., (1986). Visuomotor Fields of the Superior Colliculus: A Quantitative Model, *Vis. Res.*, 26(6): 857–873. DOI: 10.1016/0042-6989(86)90144-6

Pierre, D.A., *Optimization Theory with Application*, Wiley, New York, pp. 277–280, 1969. 96

Port, N.L., Sommer, M.A. and Wurtz, R.H., (2000). Multielectrode evidence for spreading activity across the superior colliculus movement map, *J. Neurophys.*, Jul 84(1): 344–57.

Quaia, C. and Optican, L.M., (1998). Commutative saccadic generator is sufficient to control a 3-D ocular plant with pulleys. *J. Neurophysiol.*, 79 (6), 3197–3215.

Quaia, C. and Optican, L.M., (2003). Dynamic eye plant models and the control of eye movements. *Strabismus*, 11(1), 17–31. DOI: 10.1076/stra.11.1.17.14088

Ramat, S., Leigh, R.J., Zee, D. and Optican, L., (2007). What Clinical Disorders Tell Us about the Neural Control of Saccadic Eye Movements. *Brain*, 1–26. DOI: 10.1016/j.ajo.2007.01.007

Ramat, S., Leigh, R.J., Zee, D. and Optican, L., (2005). Ocular Oscillations generated by Coupling of Brainstem Excitatory and Inhibitory Saccadic Burst Neurons. *Exp. Brain Res.*, 160: 89–106. DOI: 10.1007/s00221-004-1989-8

Raybourn, M.S. and Keller, E.L., (1977). Colliculoreticular organization in primate oculomotor system. *J. Neurophysiol.*, 40: 861–878.

Ricardo, J.A., (1981). Efferent connections of the subthalamic region in the rat. II. The zona incerta, *Brain Res.*, Jun 9; 214(1): 43–60. DOI: 10.1016/0006-8993(81)90437-6

Robinson, D.A., Models of mechanics of eye movements. In: B.L. Zuber (Ed.), *Models of Oculomotor Behavior and Control* (pp. 21–41). Boca Raton, FL: CRC Press, 1981. 77, 78, 82, 93, 102, 128

Robinson, D.A., (1973). Models of the saccadic eye movement control system. *Kybernetik*, 14: 71–83. DOI: 10.1007/BF00288906 15

Robinson, D.A., (1972). Eye movements evoked by collicular stimulation in the alert monkey, *Vis. Res.*, 12: 1795–1808. DOI: 10.1016/0042-6989(72)90070-3

Robinson, D.A., O'Meara, D.M., Scott, A.B. and Collins, C.C., (1969). Mechanical components of human eye movements, *J. Appl. Physiol.* 26, 548–553. 77, 78, 79

Robinson, D.A., (1964). The Mechanics of Human Saccadic Eye Movement, *J. Physiol.,* London, 174: 245.

Sato, H. and Noda, H., (1992). Saccadic dysmetria induced by transient functional decoration of the cerebellar vermis, *Brain Res. Rev.,* 8(2): 455–458. DOI: 10.1007/BF02259122

Sato, Y. and Kawasaki, T., (1990). Operational unit responsible for plane-specific control of eye movement by cerebellar flocculus in cat, *J. Neurophys.,* 64(2): 551–564.

Schweighofer, N., Arbib, M.A. and Dominey, P.F., (1996). A model of the cerebellum in adaptive control of saccadic gain. II. Simulation results, *Biol. Cyber.,* Jul 75(1): 29–36. DOI: 10.1007/BF00238737

Scudder, C.A., Kaneko, C., Fuchs, A., (2002). The brainstem burst generator for saccadic eye movements: a modern synthesis. *Exp. Brain Res.,* 142: 439–462. DOI: 10.1007/s00221-001-0912-9

Seidel, R.C., (1975). Transfer-function-parameter estimation from frequency-response data ˜ a FORTRAN program, NASA TM X-3286 Report. 74

Short, S.J. and Enderle, J.D., (2001). A Model of the Internal Control System Within the Superior Colliculus, *Biomed. Sci. Instru.,* 37: 349–354.

Smith, Jr., F.W., (1961). Time-optimal control of higher-order systems, *IRE Trans. Automat. Contr.,* vol. AC-6, pp. 16–21. 96

Sparks, D.L., (2002). The Brainstem Control of Saccadice Eye Movements. *Neuroscience,* 3: 952–964.

Sparks, D.L. and Hartwich-Young, R., (1989). The Deep Layers of the Superior Colliculus. *Rev. Oculomot. Res.,* 3: 213–55.

Sparks, D.L. and Nelson, J.S., (1987). Sensory and motor maps in mammalian superior colliculus. *TINS,* vol. 10, no. 8: 312–317. DOI: 10.1016/0166-2236(87)90085-3

Sparks, D.L., (1986). Translation of Sensory Signals Into Commands for Control of Saccadic Eye Movements: Role of Primate Superior Colliculus, *Physiol. Rev.,* Jan 66(1): 118–171.

Sparks, D.L., (1978). Functional Properties of Neurons in the Monkey Superior Colliculus: Coupling of Neuronal Activity and Saccade Onset, *Brain Res.,* 156: 1–16. DOI: 10.1016/0006-8993(78)90075-6

Sparks, D.L., Holland, R. and Guthrie, B.L., (1976). Size and Distribution of Movement Fields in the Monkey Superior Colliculus, *Brain Res.,* 113: 21–34. DOI: 10.1016/0006-8993(76)90003-2

Stanton, G.B., Goldberg, M.E. and Bruce C., (1988). Frontal eye field efferents in the macaque monkey. I. Subcortical pathways and topography of striatal and thalamic terminal fields. *J. Comp. Neurol.*, 271: 473–492. DOI: 10.1002/cne.902710402

Stechison, M.T., Saint-Cyr, J.A. and Spence, S.J., (1985). Projections from the nuclei prepositus hypoglossi and intercalatus to the superior colliculus in the cat: an anatomical study using WGA-HRP. *Experimental Brain Research*, vol. 59, no. 1: 139–50. DOI: 10.1007/BF00237674

Sylvestre, P.A. and Cullen, K.E., (2006). Premotor Correlates of Integrated Feedback Control for Eye–Head Gaze Shifts. *J. Neurophysiol.*, 26(18):4922–4929. DOI: 10.1523/JNEUROSCI.4099-05.2006

Sylvestre, P.A. and Cullen, K.E., (1999). Quantitative analysis of abducens neuron discharge dynamics during saccadic and slow eye movements. *J. Neurophysiol.*, 82 (5), 2612–2632.

Takagi, M., Zee, D.S. and Tamargo, R.J., (1998). Effects of lesions of the oculomotor vermis on eye movements in primate: saccades, *J. Neurophysiology*, Oct 80(4): 1911–1931.

Van Gisbergen, J.A., Robinson, D.A. and Gielen, S. (1981). A quantitative analysis of generation of saccadic eye movements by burst neurons. *J. Neurophysiol.*, 45 (3), 417–442.

Van Opstal, A.J., Van Gisbergen, J.A.M. and Eggermont, J.J., (1985). Reconstruction of Neural Control Signals for Saccades Based on a Inverse Method, *Vis. Res.*, 25(6): 789–801. DOI: 10.1016/0042-6989(85)90187-7

Versino, M., Hurko, O. and Zee, D.S., (1996). Disorders of binocular control of eye movements in patients with cerebellar dysfunction, *Brain*, Dec 119 (Pt 6): 1933–1950. DOI: 10.1093/brain/119.6.1933

Vilis, T., Snow, R. and Hore, J., (1983). Cerebellar saccadic dysmetria is not equal in the two eyes, *Exp. Brain Res.*, 51: 343–350. DOI: 10.1007/BF00237871

Vossius, G., (1960). The system of eye movement. *Z. Biol.*, vol. 112: 27–57.

Weber, R.B. and Daroff, R.B., (1972). Corrective movements following refixation saccades: Type and control system analysis. *Vision Res.*, 12:467–475.

Westheimer, G., (1954). Mechanism of saccadic eye movements. *AMA Archives of Ophthalmology*, 52: 710–724. DOI: 10.1016/0042-6989(72)90090-9

Westine, D.M. and Enderle, J.D., (1988). Observations on neural activity at the end of saccadic eye movements. *Biomedical Sciences Instrumentation*, 24: 175–185.

Widrick, J.J., Romatowski, J.G., Karhanek, M. and Fitts, R.H., (1997). Contractile properties of rat, rhesus monkey, and human type I muscle fibers. *Am. J. Physiol.*, 272, R34–42.

Wolfe, J.W., Engelken, E.J., Olson, J.W. and Allen, J.P., (1978). Cross-power spectral density analysis of pursuit tracking. *Annals of Otology, Rhinology and Laryngology*, 87(6): 837–844. 9

Wurtz, R.H. and Goldberg, M.E., (1972). Activity of superior colliculus in behaving monkey. III. Cells discharging before eye movements. *Journal of Neurophysiology*, vol. 35: 575–586.

Yasui, S. and Young, L.R., (1975.) Perceived visual motion as effective stimulus to pursuit eye movement system. *Science*, 190: 906–908. 21

Zee, D.S., Optican, L.M., Cook, J.D., Robinson, D.A. and Engel, W.K., (1976). Slow saccades in spinocerebellar degeneration, *Arch. Neurol.*, vol. 33, pp. 243–251. DOI: 10.1126/science.1188373 92

Zhou, W., Chen, X. and Enderle (2009). An Updated Time-Optimal 3rd-Order Linear Saccadic Eye Plant Model. *International Journal of Neural Systems*, vol. 19, no. 5: 309–330, 2009. 4

Author's Biography

JOHN D. ENDERLE

John D. Enderle, Biomedical Engineering Program Director and Professor of Electrical & Computer Engineering at the University of Connecticut, received the B.S., M.E., and Ph.D. degrees in biomedical engineering, and M.E. degree in electrical engineering from Rensselaer Polytechnic Institute, Troy, New York, in 1975, 1977, 1980, and 1978, respectively.

Dr. Enderle is a Fellow of the IEEE, the past Editor-in-Chief of the *EMB Magazine* (2002-2008), the 2004 EMBS Service Award Recipient, Past-President of the IEEE-EMBS, and EMBS Conference Chair for the 22nd Annual International Conference of the IEEE EMBS and World Congress on Medical Physics and Biomedical Engineering in 2000. He is also a Fellow of the American Institute for Medical and Biological Engineering (AIMBE), Fellow of the American Society for Engineering Education and a Fellow of the Biomedical Engineering Society. Enderle is a former member of the ABET Engineering Accreditation Commission (2004-2009). In 2007, Enderle received the ASEE National Fred Merryfield Design Award. He is also a Teaching Fellow at the University of Connecticut since 1998.

Enderle is also involved with research to aid persons with disabilities. He is Editor of the NSF Book Series on *NSF Engineering Senior Design Projects to Aid Persons with Disabilities*, published annually since 1989. Enderle is also an author of the book *Introduction to Biomedical Engineering*, published by Elsevier in 2000 (first edition) and 2005 (second edition). Enderle's current research interest involves characterizing the neurosensory control of the human visual and auditory system.

Printed in the United States
by Baker & Taylor Publisher Services